Electronic Measurements
A Practical Approach

Synthesis Lectures on Electrical Engineering

Electronic Measurements: A Practical Approach

Farzin Asadi and Kei Eguchi

ISBN: 978-3-031-00893-1   paperback
ISBN: 978-3-031-02021-6   ebook
ISBN: 978-3-031-00136-9   hardcover

DOI 10.1007/978-3-031-02021-6

A Publication in the Springer series
SYNTHESIS LECTURES ON ELECTRICAL ENGINEERING

Lecture #7
Series Editor: Richard C. Dorf, *University of California, Davis*
Series ISSN
Print 1559-811X   Electronic 1559-8128

Electronic Measurements

A Practical Approach

Farzin Asadi
Maltepe University, Istanbul, Turkey

Kei Eguchi
Fukuoka Institute of Technology, Fukuoka, Japan

SYNTHESIS LECTURES ON ELECTRICAL ENGINEERING #7

ABSTRACT

Measurement is the process of obtaining the magnitude of a quantity relative to an agreed standard. Electronic measurement, which is the subject of this book, is the measurement of electronic quantities like voltage, current, resistance, inductance, and capacitance, to name a few.

This book provides practical information concerning the techniques in electronic measurements and knowledge on how to use the electronic measuring instruments appropriately. The book is composed of five chapters.

Chapter 1 focuses on digital multimeters. You will learn how to use it for measurement of AC/DC voltages/currents, resistance, connection test, and diode forward voltage drop test.

Chapter 2 focuses on power supplies. Although power supplies are not a measurement device, they have an undeniable role in many measurements. So, being able to use power supplies correctly is quite important.

Chapter 3 focuses on function generators. Like the power supplies, the function generators are not a measurement device in the first look. However, they play a very important role in many electronic measurements. So, being able to use a function generator correctly is an important skill any technician or engineer needs.

Chapter 4 focuses on oscilloscopes. These days, digital oscilloscopes are the most commonly used tool in both industry and university. Because of this, this chapter focuses on digital oscilloscopes not on the analog ones which are almost obsolete.

Chapter 5 focuses on drawing graph of data you obtained from your measurement. Visualization of data is very important in practical works. This chapter show how you can use MATLAB® for drawing the graph of your measurements.

This book could be used a laboratory supplement for students of electrical/mechanical/mechatronics engineering, for technicians in the field of electrical/electronics engineering, and for anyone who is interested to make electronic circuits.

KEYWORDS

capacitor, electronic measurement, digital multimeter, digital storage oscilloscope, function generator, inductor, oscilloscope, power supply, resistor

Contents

Preface

"When you can measure what you are speaking about, and express it in numbers, you know something about it; but when you cannot measure it, when you cannot express it in numbers, your knowledge is of a meagre and unsatisfactory kind."

Lord Kelvin

Measurement is the process of obtaining the magnitude of a quantity relative to an agreed standard. Measurement is fundamental to the sciences, engineering, construction, and other technical fields, and to almost all everyday activities. The instruments used to measure any quantity are known as measuring instruments.

Electronic measurement, which is the subject of this book, is the measurement of electronic quantities like voltage, current, resistance, inductance, and capacitance, to name a few.

This book provides practical information concerning the techniques in electronic measurements and knowledge on how to use the electronic measuring instruments appropriately. This book could be used a laboratory supplement for students of electrical/mechanical/mechatronics engineering, for technicians in the field of electrical/electronics engineering, and for anyone who is interested to make electronic circuits.

This book contains five chapters.

Chapter 1 focuses on digital multimeters. A digital multimeter is a test tool used to measure two or more electrical values, principally voltage, current, and resistance. It is a standard diagnostic tool for engineers and technicians in the electrical/electronic industries. You will learn how to use it for measurement of AC/DC voltages/currents, resistance, connection test, and diode-forward voltage drop test.

Chapter 2 focuses on power supplies. A power supply is an electrical device that supplies electric power to an electrical load. Power supplies convert high AC voltage of the grid into a (usually) 0-30 V DC range. Power supply is not a measurement device. However, it is an important component of all the electronic laboratories and has an undeniable role in many measurements.

For instance, assume that you have a cable and you want to measure the resistance of it. Since the resistance is very low, you cannot use the digital multimeter. Because the digital multimeters cannot measure very low values of resistance and simply show 0 for very low values. In this case, you can use a power supply to pass a known value of current from the cable and you use a digital multimeter to measure the voltage drop across the cable. Using the Ohm's law, you can find the cable resistance easily. So, being able to use power supply correctly is quite important.

Chapter 3 focuses on function generators. Function generators are a form of test instrument that can generate waveforms with common shapes: sine, square, pulse, triangular, sawtooth, etc. Like the power supplies, the function generators are not a measurement device in the first look. However, they play a very important role in many electronic measurements.

For instance, assume that you want to measure capacitance of a capacitor however you don't have a capacitance meter. In this case, you can make a simple series RC circuit (with a known resistor) and stimulate it with a sinusoidal waveform. Based on the magnitude of the capacitor voltage or phase difference between the capacitor voltage and stimulation point voltage, you can calculate the value of capacitor. So, being able to use function generator correctly is quite important.

Chapter 4 focuses on oscilloscopes. An oscilloscope or scope is a type of electronic measuring instruments that graphically displays varying signal voltages, usually as a calibrated two-dimensional plot of one or more signals as a function of time. Oscilloscopes could be divided into two broad categories: analog oscilloscopes and digital storage oscilloscopes (DSO). The difference between an analog oscilloscope and a digital oscilloscope is that in an analog device the waveform is shown in the original form, while a digital oscilloscope converts the original analog waveform by sampling it and converts them into digital numbers and then stores them in digital format. This is done by an Analog-to-Digital (A/D) converter. Digital technology progress made the DSOs the most common type of oscilloscope in use. DSOs provide advanced trigger, storage capability and automatic measurement for the user. And all of them come with a low price. Because of this, this chapter focuses on digital oscilloscopes not on the analog ones which are almost obsolete.

Chapter 5 focuses on drawing graph of data you obtained from your measurement. Visualization of data is very important in practical works. Remember that one figure is worth a 1,000 words. This chapter shows how you can use MATLAB® for drawing the graph of your measurements.

Measurement devices which are used in electrical and electronics engineering is not limited to the devices which listed above, however the listed devices are the most fundamental ones that every technician and engineer must know.

Electronic measurement devices are produced by many different companies. It is impossible to show the details of all the available measurement devices in the market. In fact, such a thing is not necessary at all!

If you are able to work with a measurement device which is produced by company A, then certainly you can work with a measurement device which is made by company B. The only thing that you need is to take a look to the user manual of the new device.

In this book, we tried to teach how to fish instead of giving fish to you. We tried to teach the logic behind the electronic measurement techniques instead of involving with the details of a special device. So, if you want to use a measurement device that is not studied in this book, don't be worried at all. After reading the book, simply take a look at the user manual of your device in order to see its specific details.

We hope that this book will be useful for its readers, and we welcome comments on the book.

Farzin Asadi and Kei Eguchi
farzinasadi@maltepe.edu.tr and eguti@fit.ac.jp
February 2021

CHAPTER 1

Digital Mutimeter

1.1 INTRODUCTION

Multimeters are one of the most commonly used measurement devices. They can measure resistance, voltage (AC or DC), current (AC or DC), and in some models temperature, frequency, voltage drop of diodes, and capacitance. Multimeters could be divided into two categories: analog multimeters and Digital Multi Meters (DMM).

Analog multimeters have a moving pointer and a scale in order to show the measurement result (Fig. 1.1). The digital multimeters don't have any moving parts, but rather a display in order to show the result. Analog multimeters are obsolete today and because of that we only study the DMMs.

DMMs have different types. Some types are designed to be portable (Fig. 1.2). These types are supplied from a battery and are lightweight. Another commonly used type is the desktop multimeter (Fig. 1.3) which are designed to be supplied from city electricity.

The price of DMMs depend on the accuracy of the device and vary from a few dollars up to thousands of dollars. Needless to say, the more accurate devices are more expensive.

Figure 1.1: An analog multimeter.

Figure 1.2: Portable DMM.

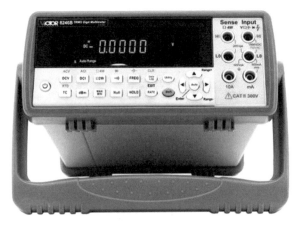

Figure 1.3: Desktop DMM. Typically, desktop DMMs are supplied from the city electricity.

Figure 1.4: A switched DMM.

The DMMs could be divided into two groups: "Switched" multimeters and "Auto-Range" multimeters. In switched multimeters, selection of the suitable range for measurement is the duty of user, however in auto-range multimeters, this job is done automatically by the device itself. Figures 1.4 and 1.5 show the switched and auto-range multimeters, respectively.

Assume that you want to measure the voltage of a car battery. If you do this measurement with a switched DMM, then you need to put the selector in the DC voltage measurement section, and select a suitable range based on the value which you want to measure. We know that a nominal value of a car battery is 12 V DC. So, the 20 V range is a good option for this measurement because the measured value is less than 20 V. Note that you can use 200 V or 500 V ranges for this measurement as well. However, your measurement will not be precise. The maximum precision is given by the 20 V range. For instance, when the 20 V range is selected, you read 11.3 V. However, when you use the 200 V or 500 V range, you may see 11 or 12 V on the display.

1.2   WORKING PRINCIPLE OF DMM

In this section we will study the working principle of DMM. The heart of a DMM is an Analog to Digital Converter (ADC). The DMM needs to convert resistance or current signals into voltage signal before being able to read it with the aid of ADC. Assume we want to measure

Figure 1.5: An auto-range DMM.

the value of a resistor. If we apply a known current to the resistor, then the value of voltage drop across the resistor depends on the resistor value according to Ohm's law ($R = \frac{V}{I}$). So, if we divide the voltage drop to the known value of current, then we could obtain the resistance. In the same way, we can convert a current signal into a voltage signal. For instance, assume we have an unknown current and we want to measure it. In this case, we could use a simple low value resistor (with known value) as a sensor and measure the voltage drop across it. The voltage drop divided by the known value of sense resistor give us the value of unknown current. Hall effect sensors could be used as another method for conversion of a current signal into a voltage signal.

Now assume that we want to measure the Root Mean Square (RMS) of sinusoidal signal shown in Fig. 1.6. In this case we could use the peak detector circuit shown in Fig. 1.7. If we neglect the voltage drop of the diode, then the capacitor will be charged up to the peak value of the AC signal. If we divide the peak value by $\sqrt{2} = 1.41$, then we will obtain the RMS value of sinusoidal signal.

Note that this technique is applicable if the input is sinusoidal. RMS of the waveforms shown in Fig. 1.8 could not be measured in this way.

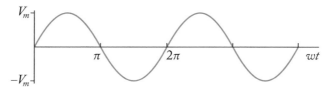

Figure 1.6: Purely sinusoidal waveform.

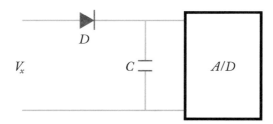

Figure 1.7: Simple peak detector.

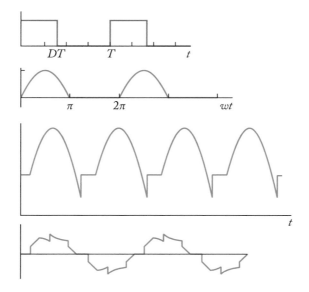

Figure 1.8: Non-sinusoidal waveform.

If you need to measure the RMS of waveforms shown in Fig. 1.8, then you need to sample them with a high enough frequency and then use the discrete form of the RMS formula in order to calculate the RMS (Fig. 1.9).

DMMs which are able to do such high-frequency sampling and calculation tasks are called "True RMS" DMMs. When you see the "True RMS" label on a DMM (see Fig. 1.10), then you

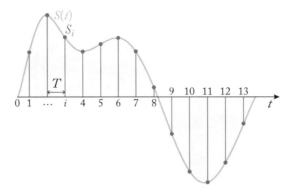

Figure 1.9: Sampling an analog waveform.

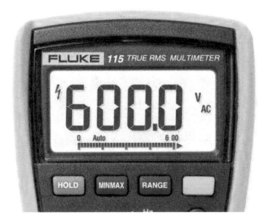

Figure 1.10: True RMS type DMM.

can measure RMS of none sinusoidal waveforms. Generally, the price of "True RMS" DMM's are higher than none true RMS ones. None true RMS ones could be used for measurement of RMS of sinusoidal waveforms only.

True RMS DMM's could be divided into two groups: True RMS AC and True RMS AC + DC. The difference could be understood with the aid of a numeric example. Assume

$$v(t) = 5 + 10\sin(2\pi \times 50 \times t).$$

From a mathematical point of view, RMS of

$$V_0 + \sum_{i=1,2,3,\dots} V_i \sin(i \times \omega_0 t + \varphi_i)$$

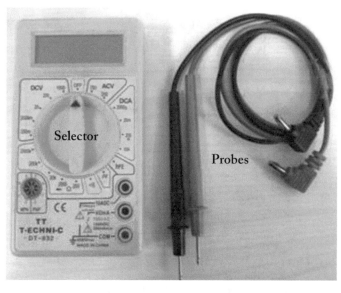

Figure 1.11: A cheap DMM.

is

$$\sqrt{V_0^2 + \frac{1}{2}(\sum_{i=1,2,3,\ldots} V_i^2)}.$$

So the RMS of the given $v(t)$ is

$$\sqrt{5^2 + \frac{1}{2} \times 10^2} = 8.66 \text{ V}.$$

Assume we applied $v(t) = 5 + 10\sin(2\pi \times 50 \times t)$ to a True RMS AC DMM. Display of such a DMM will show $\frac{10}{\sqrt{2}} = 7.07$ V. In other words, the True RMS AC DMMs consider the AC component (the $10\sin(2\pi \times 50 \times t)$ term) of input signal only. Remember that the RMS of a sinusoidal signal is the peak value divided by $\sqrt{2}$.

If you apply the given $v(t)$ to a True RMS AC + DC DMM, then you will see 8.66 V on the display. In other words, the True RMS AC + DC consider both the AC and DC components of the signal to measure the RMS.

1.3   MEASUREMENT WITH DMM

A cheap typical switched DMM is shown in Fig. 1.11. It comes with two probes. The probes are responsible to transfer the signal under measurement to the internal circuit of DMM without perturbing it. In fact, they are nothing more than a piece of wire with low resistance, i.e., 150 mΩ.

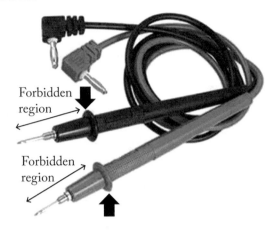

Figure 1.12: DMM probes.

Figure 1.13: Correct way of holding the probes.

The maximum voltage/current that can be measured safely with the probes are written on the probe body. Never try to measure values that are bigger than those values.

The DMM probes have a ring on them in order to keep the user hand far from the metal part of the probe which is in direct contact with the circuit under test. Never enter your fingers into the forbidden region of the probe (see Fig. 1.12). Always hold the probes as shown in Fig. 1.13. Never hold the probes as shown in Fig. 1.14. Otherwise, expect an electric shock or even death!

Figure 1.14: Never touch the metal part of the probes.

One of the probes is red and the other one is black. You will see the difference between these two probes in the next sections. This DMM is not a true RMS DMM. So, we can measure only the RMS of pure sinusoidal signals correctly.

With the aid of cheap DMM shown in Fig. 1.11, we could measure:

1. AC/DC voltages;

2. DC currents (this cheap DMM could not measure AC currents);

3. transistor current gain $\beta = h_{FE} = \frac{I_C}{I_B}$ and forward voltage drop of diode;

4. short circuit test;

5. resistance; and

6. frequency.

In the next sections, we will see how we can obtain do these measurements.

1.4 DMM JACKS

The DMM in Fig. 1.15 has three jacks. The black probe is always connected to the COM jack, but the red probe must be connected to one of the remaining two jacks. The jack labeled with "VΩmA" is used for measurement of voltage (DC or sinusoidal), resistance, and current in the mA range. You put the red probe in the jack with "VΩmA" label when you want to measure the

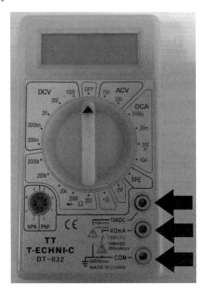

Figure 1.15: Jacks of DMM.

forward voltage drop of diode and the short circuit test as well. The other jack is labeled as 10 ADC and is used for DC currents more than 200 mA up to 10 A.

1.5   MEASUREMENT OF DC VOLTAGES

In order to measure a DC voltage,

1. connect the black probe to COM jack (Fig. 1.16);

2. connect the red probe to "V Ω mA" jack (Fig. 1.17); and

3. put the selector in the section labeled with DCV (Fig. 1.18). Connect the probes to the points whose voltage difference you want to measure and read the number shown on the display of DMM.

The DCV section of switched DMM are labeled with numbers that show the maximum value of quantity that is measurable in that state. For instance, in this DMM, the DCV section labels are "200 m, 2000 m, 20, 200, 1000." If you put the selector in the 200 m mode, then you can measure up to 200 mV and if you put in the 200 section, then you can measure up to 200 VDC. When you don't have any idea about the maximum of voltage under measurement, then select the maximum range, i.e., 1000 V (Fig. 1.19). After selection of maximum range, do the measurement and obtain an approximation about the value of voltage under measurement. For instance, assume that your reading is 12 V. After this measurement, you have an idea about

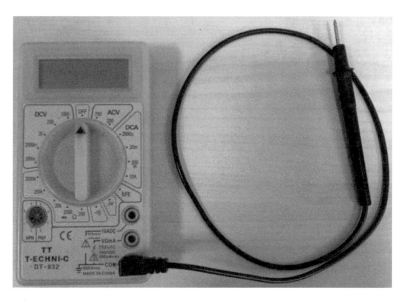

Figure 1.16: Black probe is always connected to the COM jack.

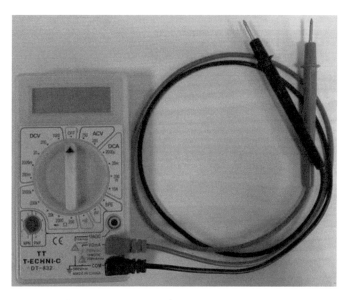

Figure 1.17: Connect the red probe to V Ω mA.

Figure 1.18: Put the selector in the DCV section.

Figure 1.19: Selector is in the 1000 V mode. Reading is 12 V.

Figure 1.20: Selector is in the 20 V mode. Reading is 11.84 V.

the value of signal. After this step, you can select a better mode in order to have a more accurate reading. For instance, you can put your selector in the 20 V section and see that the accurate value of signal under measurement is something like 11.84 V (Fig. 1.20).

You can put your selector in the 200 section as well for this measurement since $12 < 200$. However, as you see in Fig. 1.21, in this case you read 11.8 V which has one significant digit less than the value which you read with the 20 V range.

Note 1:   The voltage shown on the display of the DMM is the voltage difference between the red probe and black probe. For instance, assume that you see 11.84 V on the display of DMM. This means that the voltage of the point which red probe is connected to is 11.84 V higher than the voltage of the point which the black probe is connected to.

Sometimes the value shown on the display is negative. For instance, if you connect the red probe to the negative terminal of a small battery and the black probe to the positive terminal of the battery, then the value shown on the display of DMM is negative.

When the shown value is negative, the potential of the point which red probe is connected to is less than the potential of the point which the black probe is connected to.

Note 2:   When you want to measure the voltage, you need to connect the volt meter in parallel with the load (Fig. 1.22).

Figure 1.21: Selector is in the 200 V mode. Reading is 11.8 V.

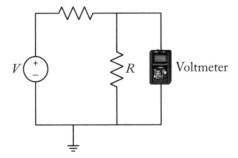

Figure 1.22: Voltmeter must be connected in parallel to the load.

Note 3: When the input signal is periodic, the value shown on the display is the average value of the signal. The average value of periodic function $f(t) = f(t + T)$ is defined as:

$$\overline{f(t)} = \frac{1}{T} \int_{t_0}^{t_0+T} f(t) \, dt.$$

For instance, if you apply the saw tooth signal shown in Fig. 1.23 to a DMM which is in DC voltage measurement mode; then you will read

$$\overline{f(t)} = \frac{1}{T} \int_{t_0}^{t_0+T} f(t) \, dt = \frac{1}{T} \times \left(\frac{1}{2} \times V \times T \right) = \frac{V}{2}$$

on the display. V shows the amplitude of the saw tooth signal applied to the DMM.

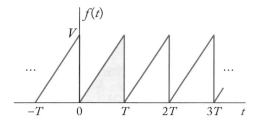

Figure 1.23: A saw tooth signal with amplitude V.

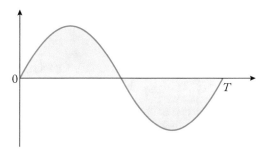

Figure 1.24: Integral of one period of a sinusoidal waveform is zero.

If you apply a purely sinusoidal signal $\left(f(t) = V_m \sin\left(\frac{2\pi}{T}t\right)\right)$ to the DMM in the DC voltage measurement mode, the display shows zero since the average value of purely sinusoidal signal is zero. The reason is obvious: the area of the positive half cycle is equal to the area of negative half cycle, so the integral of one cycle will be zero and this lead to zero average value (Fig. 1.24).

If you apply the signal shown in Fig. 1.25 to a DMM in the DC voltage measurement mode, the reading will be:

$$\frac{1}{T} \int_0^T V_0 + V_m \sin\left(\frac{2\pi}{T}t\right) dt = \frac{1}{T} \int_0^T V_0 dt + \frac{1}{T} \int_0^T V_m \sin\left(\frac{2\pi}{T}t\right) dt$$

$$= \frac{1}{T} \times V_0 \times T + 0 = V_0.$$

For the signal shown in Fig. 1.26, the DMM reading will be

$$\frac{1}{T} \times V_{High} \times \Delta T_{ON} = V_{High} \times \frac{\Delta T_{ON}}{T} = V_{High} \times D.$$

$D = \frac{\Delta T_{ON}}{T}$ is called the duty ratio of a signal.

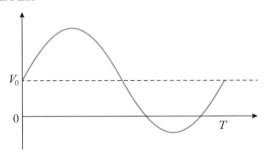

Figure 1.25: Graph of $V_{in}(t) = V_0 + V_m \sin\left(\frac{2\pi}{T}t\right)$.

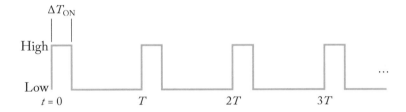

Figure 1.26: A pulse waveform.

1.6   MEASUREMENT OF AC VOLTAGES

You can measure the RMS of sinusoidal voltages with this multimeter. The procedure is the same as the measurement of DC voltages. However, instead of putting the selector in the DCV section, we need to put it in the ACV section. Like the DC measurement, the red probe is connected to the "V Ω mA" section. (As you remember, the black probe is always connected to the COM jack and only the red probe place is changed based on the type of measurement you want to do.)

When you want to measure the RMS, you never see a negative value on the display since the RMS value of any non-zero signal is always bigger than zero, i.e., it is positive.

Assume you want to measure the AC voltage between point A and B. If you connect the red probe to point A and the black probe to point B, your reading will be the same as with the case that red probe is connected to point B and black probe is connected to point A (Fig. 1.27). Remember that changing the place of probes in DC voltage measurement leads to observing a negative value.

RMS value of a sinusoidal signal is calculated as $\frac{V_m}{\sqrt{2}}$, where V_m shows the peak value of sinusoidal signal. So, if you multiply the value shown on the display, you can obtain the peak value of signal.

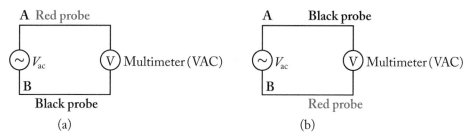

Figure 1.27: Measurement of AC voltages. You see the same number on display in both (a) and (b) cases.

Note: Any measurement device could measure in a certain frequency range. In order to obtain the frequency response of your DMM in the AC voltage measurement mode, connect the DMM probes to the output of a function generator. Select the sinusoidal output waveform. Set the frequency to 50 Hz and increase the amplitude knob of the signal generator until you see 1 Vrms on the display. Now increase the output frequency of the function generator until the value on the DMM display decreases to $\frac{1}{\sqrt{2}} = 0.707$ Vrms. The frequency which corresponds to this value is the -3 dB of your measurement device.

1.7   MEASUREMENT OF DC CURRENTS

If you want to measure the DC currents,

1. put the red probe in the suitable jack (see Figs. 1.28–1.30);

2. put the mode selector in the DC current section and select the suitable range. If you don't have any information about the range of current signal that you want to measure, then select the maximum range which generally is 10 A (Fig. 1.31); and

3. put the multimeter in series with the circuit that you want to measure its current (Fig. 1.32).

   Note that the number shown on the DMM's display is positive if the current under measurement enters the red probe and exit from the black probe (Fig. 1.33). When the current enters the black probe and exit from the red probe, the number shown is negative (Fig. 1.34).

   If you apply a periodic signal to a DMM in the DC current measurement mode, it shows the average value of the signal. For instance if you give a current like the one shown in Fig. 1.35, the read value will be $\frac{6 \times 5 \text{ m}}{25 \text{ m}} = 1.2$ A.

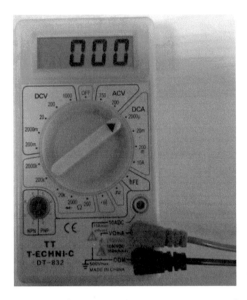

Figure 1.28: The COM and V Ω mA is used for measurement of currents up to 2000 μA = 2 mA. The mode selector is in the 2000 μA state.

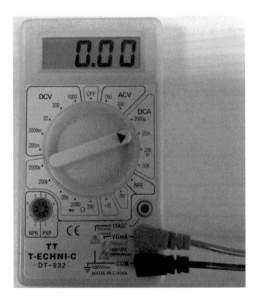

Figure 1.29: The COM and V Ω mA is used for measurement of currents up to 20 mA. The mode selector is in the 20 mA state.

Figure 1.30: The COM and V Ω mA is used for measurement of currents up to 200 mA. The mode selector is in the 200 mA state.

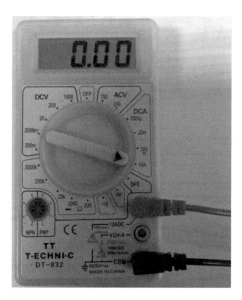

Figure 1.31: The COM and V Ω mA is used for measurement of currents up to 10 A. The mode selector is in the 10 A state.

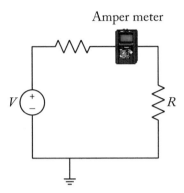

Figure 1.32: In order to measure the current, the ammeter must be connected in series with the load.

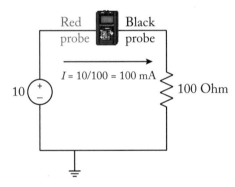

Figure 1.33: Display of DMM will show +100 mA.

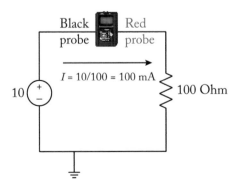

Figure 1.34: Display of DMM will show −100 mA.

Figure 1.35: Pulsating current with frequency of 40 Hz.

1.8   AC CURRENT MEASUREMENT

The procedure to measurement AC currents is quite similar to DC currents. The cheap DMM shown in Fig. 1.11 can't measure the AC currents. However, in models which are able to measure AC currents the following is true.

1. Red probe is connected to the suitable jack.

2. Put the mode selector in the AC current section and select the suitable range. If you don't have any information about the range of current signal that you want to measure, then select the maximum range which generally is 10 A.

3. The value which is shown on the display of the DMM is the RMS of the current under measurement. If the DMM has the true RMS label, then it could measure the RMS of non-sinusoidal currents as well. If it is not true RMS, then you can measure the RMS of purely sinusoidal currents only.

One way to measure AC currents with the DMM shown in Fig. 1.11 is to use a small sense resistor: connect a small resistor in series with the load that you want to measure its current and measure the RMS of the voltage drop across the sense resistor. By using Ohm's law, you can calculate the RMS of current flow in the circuit. For instance, assume that the sense resistor value is 1 Ω and the voltage drop across the resistor is 0.7 Vrms. Then the RMS of current will be $I_{rms} = \frac{V_{rms}}{R} = \frac{0.7}{1} = 0.7\ A_{rms}$.

Note: If you try to measure voltage by mistake when the red probe is in the current measurement jack, then you make a short circuit and the internal fuse of DMM will be blown and you need to replace it. So, it is a good idea to take out the red probe from the current jack and return it to the voltage measurement jack immediately after your current measurement is finished. This is true for both AC and DC current measurements.

1.9   MEASUREMENT OF RESISTANCE

In order to measure resistance:

1. Red probe is connected to suitable jack, i.e., the jack which is labeled with "VΩmA."

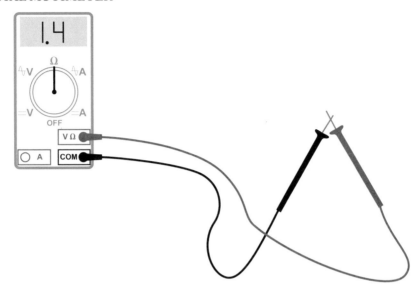

Figure 1.36: When you want to measure small resistors (i.e., in the few Ohms range), you need to know the DMM offset.

2. Mode selector is put in the resistance measurement section. The numbers which are shown in the resistance measurement section demonstrate the maximum measurable resistance. For instance, in the 20 K section, you can measure resistors up to 20 kΩ.

3. Probes are connected to the resistor whose resistance we want to measure. The number shown on the display is the resistance of the resistor under measurement.

Note: If the value of the resistor you want to measure is in the few Ohms range (Fig. 1.36), i.e., less than 10 Ohms, select the lowest range (in the DMM of Fig. 1.11, 200 Ω range) and before measuring the resistance, connect the black and red probes together, i.e., short circuit the probes together, to see the amount of offset that your DMM has. For instance, assume that after connecting the two probes together, you see 1.4 Ω on the display. Now connect the probes to the resistor whose resistance you want to measure. Assume that after connecting the resistor, the DMM shows 13 Ω. So, we deduce that the value of resistance under measurement is $13 - 1.4 = 11.6\ \Omega$.

Note: If you want to measure the resistance of a resistor which is connected to a Printed Circuit Board (PCB), then you need to take it out and isolate at least one of its legs (Fig. 1.37). The reason for this could be to understand with the aid of a simple circuit shown in Fig. 1.38. In this circuit, four 10 kΩ resistors are connected in parallel. When you connect the probes directly to the resistor under measurement (the left-hand side resistor), you will read 2.5 kΩ. In other words, other components in the circuit affect your measurement. When you isolate at least one

Table 1.1: Standard values of resistors and their color codes

Band 1	Band 2	Band 3							
		Gold	Black	Brown	Red	Orange	Yellow	Green	Blue
Brown	Black	1R0	10R	100R	1K0	10K	100K	1M0	10M
Brown	Red	1R2	12R	120R	1K2	12K	120K	1M2	12M
Brown	Green	1R5	15R	150R	1K5	15K	150K	1M5	15M
Brown	Gray	1R8	18R	180R	1K8	18K	180K	1M8	18M
Red	Red	2R2	22R	220R	2K2	22K	220K	2M2	22M
Red	Violet	2R7	27R	270R	2K7	27K	270K	2M7	27M
Orange	Orange	3R3	33R	330R	3K3	33K	330K	3M3	33M
Orange	White	3R9	39R	390R	3K9	39K	390K	3M9	39M
Yellow	Violet	4R7	47R	470R	4K7	47K	470K	4M7	47M
Green	Blue	5R6	56R	560R	5K6	56K	560K	5M6	56M
Blue	Gray	6R8	68R	680R	6K8	68K	680K	6M8	68M
Gray	Red	8R2	82R	820R	8K2	82K	820K	8M2	82M

of the legs (see Figs. 1.39 and 1.40), you don't permit other elements in the circuit to affect your measurement.

Note: You can test the transformers and incandescent lamps with the resistance measurement section of DMMs as well (Fig. 1.41). When the transformer winding is opened, or the lamp is blown, you read infinity.

Note: Nominal values of resistors is shown with the aid of color strips (see Fig. 1.42 and Table 1.1).

Example 1.1 $2R7 = 2.7\,\Omega$, $39R = 39\,\Omega$, $1K2 = 1.2\,k\Omega$, $220K = 220\,k\Omega$, $1M8 = 1.8\,M\Omega$, $82M = 82\,M\Omega$.

1.10   DIODE FORWARD VOLTAGE DROP MEASUREMENT

When current flows from the anode to cathode of a diode, the diode will be forward biased (Fig. 1.43). In this case, a voltage drop of about 0.3–0.7 V will be produced across the diode.

Using the diode test section of a DMM you can measure the voltage drop of a diode. In this mode, the DMM forward bias the diode under test with a small current and shows the voltage drop on the display. The red probe must be connected to the anode and the black probe must be connected to the cathode of the diode during the test otherwise the diode will not be

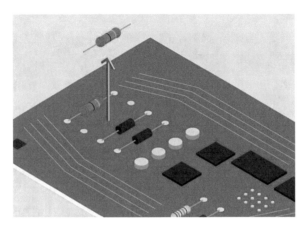

Figure 1.37: In order to have a reliable measurement, you need to isolate the part from PCB.

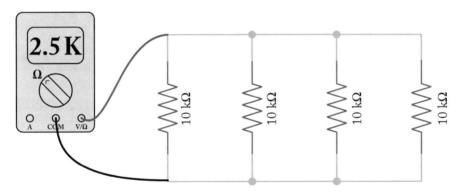

Figure 1.38: A simple parallel circuit.

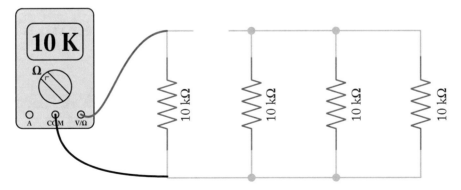

Figure 1.39: One leg of the resistor is isolated from the circuit.

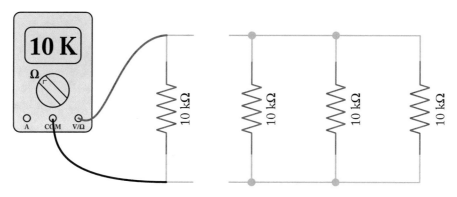

Figure 1.40: Both legs of the resistor are isolated from the circuit.

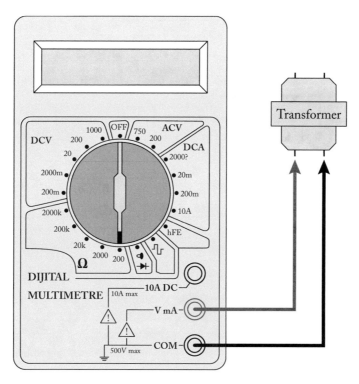

Figure 1.41: Testing a transformer with the resistance measurement section of DMM.

Color	Value
Black (2nd and 3rd bands only)	0
Brown	1
Red	2
Orange	3
Yellow	4
Green	5
Blue	6
Violet	7
Gray	8
White	9

Figure 1.42: Color codes of resistors.

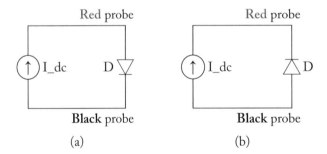

(a)   (b)

Figure 1.43: In the case (a) the diode is forward biased. In case (b) the diode is reverse biased.

forward biased and the display will show nothing. Some DMMs show the voltage drop in mV. So, when you see 700 on display, it means 700 mV or 0.7 V.

In order to measure the forward voltage drop of a diode,

1. the DMM will be put in the diode voltage drop measurement mode;

2. the red probe is put in the suitable jack. This jack is usually the same as the jack which measures the voltage and resistance;

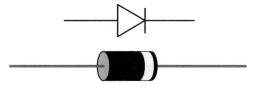

Figure 1.44: Cathode of diode is shown with a black strip.

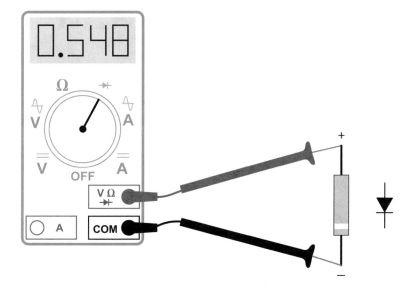

Figure 1.45: Measurement of forward voltage drop of the diode.

3. the red probe is connected to the anode of diode under test and the black probe is connected to the cathode of diode under test. The cathode of diodes generally is shown with a black strip on the body of diode (Fig. 1.44); and

4. the number shown on the display of the diode is the forward voltage drop of the diode. 0.5–0.8 V is common for Silicon diodes and 0.2–0.3 is common for germanium and Shottky diodes (Fig. 1.45).

After reading a logical forward voltage drop, connect the red probe to cathode and black probe to the anode. In this case, you must see nothing on the display of the diode otherwise the diode is faulty. Remember that a diode conducts a current only in one way; a diode which conducts a current in both directions is faulty.

Note 1:   In order to test the diodes which are mounted on PCB, you need to remove at least one of its leads from the PCB. So, the other parts of the circuit can't affect your measurement.

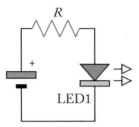

Figure 1.46: Simple test circuit for LED. Input voltage is 5 V and R is a few hundred Ohms, i.e., 220 Ω.

Note 2:   If your DMM does not have diode test capability, you can use the resistance measurement mode in order to test the diode superficially. Connect the red probe to the anode and the black probe to the cathode. You need to see a value on the display. Connect the red probe to the cathode and the black probe to the anode. In this case, you must see nothing on the display.

Note 3:   Generally, you can't use the diode test section of a DMM in order to test the Light Emitting Diodes (LED). The reason is that the DMM use a very small current in order to forward bias the diode under test. The forward voltage drop of LEDs is considerably higher than diodes. So, a DMM can't forward bias an LED and you see nothing on the display of the DMM. You can use the following simple circuit in order to test the LEDs (Fig. 1.46).

Note 4:   You can use the diode forward voltage drop measurement mode of the DMM in order to find the anode and cathode of the diode as well.

1.11   SHORT CIRCUIT TEST

You can use the DMM in order to test the continuity of wires, fuses, and PCB traces (Fig. 1.47).
   Assume that you want to find the other end of metal shown with letter A in Fig. 1.48. In order to do this, you put the DMM in the continuity test mode. This mode is generally shown with the symbol shown in Fig. 1.49. Generally, the red probe is connected to the voltage/resistance measurement jack.
   In order to do a continuity test, after selection of the continuity test mode, connect the two probes together (see Fig. 1.50) and ensure to hear the beeping sound. After hearing the beeping sound, connect one of the probes to the end that is shown with letter A in Fig. 1.48. Connect the other probe to one of the possible ends for A. When you connect the probe to the correct end, you hear the beeping sound and this shows you the other end of the wire. So, using trial and error you can find the other end of the wire.
   You can test the fuses with the continuity test mode of the DMM, as shown in Fig. 1.51. If the fuse is healthy, then you hear a beep from the multimeter (Fig. 1.52).

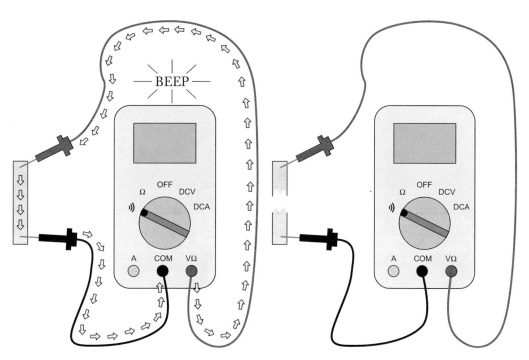

Figure 1.47: You can use the DMM in order to test the connection that potentially exists between two points.

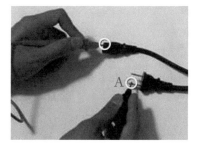

Figure 1.48: Finding the other end of point A.

Figure 1.49: Connection test icon.

Figure 1.50: Connect the probes and ensure that you hear the beeping sound.

Figure 1.51: Testing a fuse.

1.12 hFE MEASUREMENT

As you know, for a transistor operating in the linear region, the relationship between collector (I_C) and base current (I_B) is given by $I_C = \beta \times I_B$. The β shows the current gain and it is usually bigger than 100.

In order to measure the β of the transistor:

1. put the mode selector in the hFE mode;

2. enter the transistor under test into its socket (Fig. 1.53). You need to know the type of transistor under test (NPN or PNP) and its pin configuration, i.e., which pin is base, which one is the emitter, and which one is the collector. This information could be obtained using the datasheet. Note that DMM probes don't play any role in this measurementt; and

3. the β will be shown on the DMM's display.

Figure 1.52: You can use the short circuit mode of the DMM in order to see whether two points of the PCB are connected together or not.

Figure 1.53: hFE socket.

Note 1:   Current gain of the transistor is a not a constant parameter and changes with the operating point, so the number shown on the DMM is not very reliable. The DMM calculates the current gain for only one operating point and because of that we can't generalize the results of a single measurement to all operating points. Because of this, some modern DMMs do not have the hFE mode at all. However, if your DMM has this mode, you can use it as an indicator of transistor health. If you read a low number or if the display shows nothing (we assume that the transistor pins are connected to the correct hole of the socket), then the transistor may be faulty.

Figure 1.54: You cannot put this type of transistor into the hFE socket of DMM.

Note 2:   You can only fit the small transistor into the socket. If you want to test a big transistor, then you need to connect the transistor leads to the socket holes with the aid of small wires (Fig. 1.54).

Note 3:   You can use this mode in order to find the base, emitter, and collector leads of the transistor. Simply put the transistor into the socket and look at the DMM display. If it shows a number, then it means that you put the transistor leads in the correct order to the holes. In this case, the label behind the hole shows the type of pin inside the socket hole. For instance, the lead which is inside the hole with label "C" is collector. If the display shows nothing, then you need to take out the transistor and use the other replace it with a different order of pins. You continue this procedure until you see a number on the display. When you see a number on the display, this means that you put each lead in its correct place and the label behind the holes tells you the pin name.

1.13   MEASUREMENT OF INTERNAL RESISTANCES OF DMM

A DMM in the voltage/current measurement mode loads the circuit under test. The internal resistance of an ideal volt meter must be infinity in order to avoid loading the circuit under test. The internal resistance of an ideal ammeter must be zero in order to avoid loading the circuit under test. However, a real volt meter/ammeter has a finite internal resistance. In this section we show you how you can measure this value.

1.13.1   MEASUREMENT OF INTERNAL RESISTANCE OF DMM IN THE VOLTAGE MEASUREMENT MODE

Prepare a circuit like the one shown in Fig. 1.55. The voltage difference between A and B must be around $\frac{V_{dc}}{2}$ according to Ohm's law.

Now assume that we connected a DMM across points A and B in order to measure the voltage difference between A and B (Fig. 1.56).

After connecting the DMM, the resistance between points A and B will be:

$$\frac{1M \times R_x}{1M + R_x} = \frac{R_x}{1M + R_x} \times 1M \leq 1M.$$

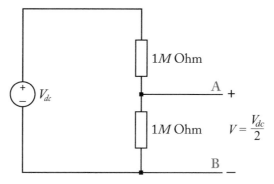

Figure 1.55: A simple voltage divider circuit.

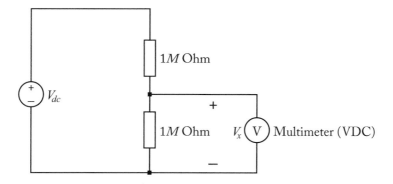

Figure 1.56: DMM is connected in parallel to the lower resistor.

Assume that DMM of Fig. 1.57 reads V_x volts, then we could write:

$$\frac{\dfrac{1M \times R_x}{1M + R_x}}{\dfrac{1M \times R_x}{1M + R_x} + 1M} \times V_{dc} = V_x.$$

We could solve this algebraic equation and calculate the value of R_x. For instance, assume $V_{dc} = 9$ V and $V_x = 3.6$ V. In this case, $R_x = 2$ MΩ. So, the internal resistance of the DMM (in the voltage measurement mode) is 2 MΩ. If you want to measure the internal resistance of the DMM in the AC voltage measurement mode, then instead of DC voltage source of Fig. 1.55, use an AC source. Calculations will not change. The only difference between the AC case and the DC case is that you use the RMS values of voltages in the calculations instead of DC values.

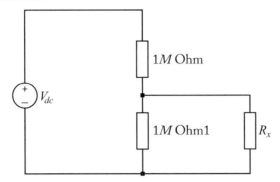

Figure 1.57: Equivalent circuit of Fig. 1.56. R_x shows the internal resistance of DMM.

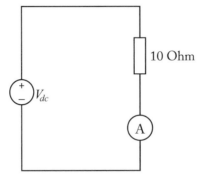

Figure 1.58: Measurement of internal resistance of ammeter. Vdc could be 2–5 V.

1.13.2 MEASUREMENT OF INTERNAL RESISTANCE OF DMM IN THE CURRENT MEASUREMENT MODE

In order to measure the internal resistance of the ammeter, set up a simple circuit like the one shown in the Fig. 1.58.

Use another DMM or an oscilloscope in order to measure the voltage drop across the ammeter (Fig. 1.59). Divide this voltage drop to the current that is shown on the display of ammeter. According to Ohm's law, the obtained number is the internal resistance of ammeter.

If you want to measure the internal resistance of an AC ammeter, then you need to measure the RMS of the voltage drop across the ammeter. This could be done with the aid of a DMM in the AC voltage measurement mode or a digital oscilloscope.

1.14 CHANGING THE BATTERY/FUSE OF DMM

Portable DMMs have a small battery which runs the internal circuit of the DMM. When the battery is consumed, you can't turn it on and you need to replace it. Before opening the DMM

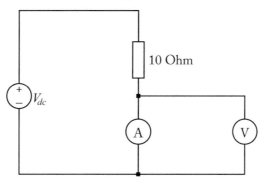

Figure 1.59: Use another DMM or an oscilloscope in order to measure the voltage drop across the ammeter. The voltage drop across the ammeter is in the mV range.

Figure 1.60: Open the back case in order to change the battery of DMM. Before opening the case, ensure that no external voltage enters the DMM circuit.

case, disconnect the probes from jacks in order to ensure that no external voltage enters the DMM circuit (Fig. 1.60). Note that large enough external voltage could shock the user and even may lead to death.

The ammeter section of DMM has a fuse protection. When the fuse is blown, you need to open the back case of DMM and replace it. Remember that when you open the back case of DMM (like the battery replacement case) you need to disconnect the probes from jacks to avoid any shock.

1.15   FURTHER READING

[1] *Electronic Measurement Systems: Theory and Practice*, 2nd ed., Institute of Physics, 1996.

CHAPTER 2

Power Supply

2.1 INTRODUCTION

All the circuits require an energy source in order to work. The power supply (PS) is responsible for providing the required energy for the circuit. The power supply takes the AC electric energy from the grid and converts it into a DC voltage. Generally, they provide the voltages in the 0–30 V range. Generally, the output current could be up to 3 A. The outputs of a PS are called a "Channel." So, when we speak about a three-channel PS (Figs. 2.1 and 2.2), we mean a PS with three outputs. Generally, the outputs are variable and the user could set them to the desired value he/she wants. Generally, PSs have one regulated output with voltage of 5 V. This output is used to supply digital circuits. Remember that traditional digital circuits work with 5 V. (However, this voltage decreased to 3.3 V and 1.1 V!) So, it is a good idea to use this fixed 5 V when you work with traditional digital circuits. You can connect a digital circuit to variable outputs of a PS. However, if you increase the voltage of that variable channel by mistake, then your circuit may be damaged. So, always use this fixed 5 V when you are working with traditional digital circuits (Fig. 2.3).

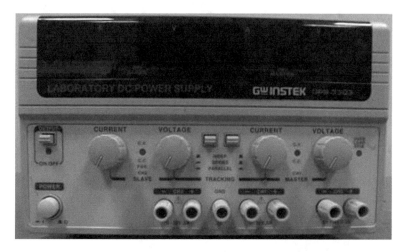

Figure 2.1: A three-channel PS. This model has two variable channels and one fixed +5 V output. Generally, the PSs are equipped with voltmeters and ammeters.

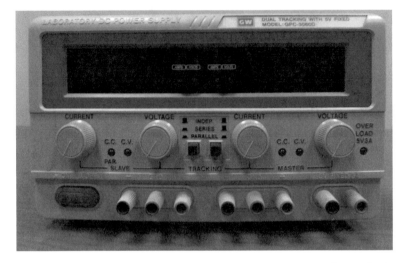

Figure 2.2: A three-channel PS. Two variable channels and one fixed (+5 V) channel.

Figure 2.3: One-channel PS.

2.2   INSIDE OF A PS

The PS uses a transformer in order to decrease the input AC voltage, for instance they decrease the 220 Vrms into 50 Vrms (Figs. 2.4 and 2.5). Then a full-wave diode rectifier and a linear regulator provides the desired output for the user.

2.3   HEAT DISSIPATIONS

When a PS converts the AC into DC, some heats are produced. These heats must get away from the PS otherwise the PS may be damaged. In order to do this, some kind of heatsink or fan must be used (Fig. 2.6). The PSs with a fan produce a little bit of audio noise when they work.

Generally, there are some switches at the back of the PS (behind the power jack) that must be set according to the voltage of the grid that you are using (Fig. 2.7).

2.4   OUTPUT CABLES OF PS

You need two wires in order to connect the output of PS to your circuit (Fig. 2.8). The wire must be thick enough to carry the current. One end of the wire has a banana plug which is connected to the PS female output. The other end of the wire has an alligator clip and is connected to the circuit under test (Figs. 2.9 and 2.10).

It is not a good idea to connect the alligator clips to the output of the PS (Fig. 2.11) because the pressure which alligator clips apply to the female jack destroys its surrounding insulator after a while.

2.5   CONTROLS OF PS

The PSs have a knob for setting the voltage and maximum allowed current. Using the voltage knob, you can determine the output voltage of channel. With the aid of current knob, you can determine the maximum allowed output current. Using the current knob, you can protect your circuit which is connected to the PS. As an example, assume that you made a small signal amplifier which draws 20 mA when it operates. Now assume something wrong happened in your circuit, for instance a short circuit. In such a case, if the power supply is capable of supplying huge amounts of currents to your circuit, then your circuit will be destroyed. Specifically, the transistors and ICs of your circuit will be burned. So, we need to limit the current to a logical value. For instance, for a circuit which require 20 mA, you can set the maximum current to 40–50 mA. 40–50 mA is big enough to supply the circuit and decrease the chance of damage in the case of a fault. So, the amount of maximum allowed current changes from circuit to circuit. If you want to supply a DC motor or a power amplifier, then you need to set the maximum allowed current in the Amper range. In fact, with the aid of current knob, you can determine the maximum power that circuit under test can absorb from the PS. For instance, if you set the

Figure 2.4: Inside of a PS.

Figure 2.5: Laboratory PS using a big transformer in order to decrease the grid voltage.

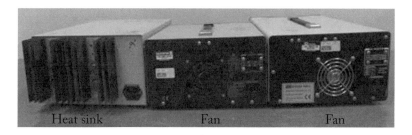

Figure 2.6: Use of fan and heatsink for decreasing the temperature of power transistors of the PS.

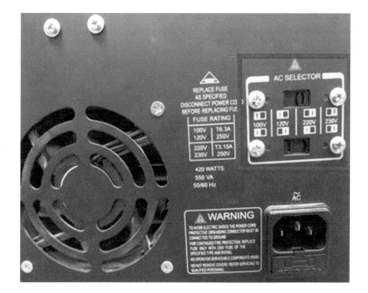

Figure 2.7: The AC selector switch must be set according to the grid voltage.

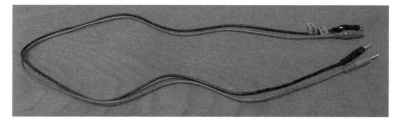

Figure 2.8: PS output cable.

Figure 2.9: PS output cables have a banana plug at one end and alligator clips at the other end.

Figure 2.10: Connection of the output cable to the PS. The output jack of the PS is designed for banana plugs.

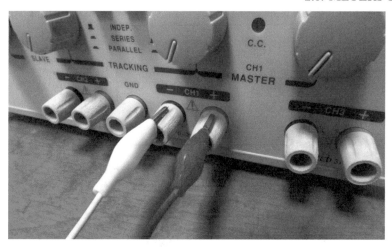

Figure 2.11: Using alligator clips for connection of the PS to the external circuit is NOT recommended.

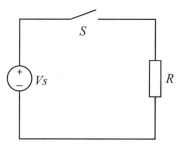

Figure 2.12: Switch S connects/disconnects the load from the source.

current knob to 500 mA and voltage knob to 9 V, then you permit the circuit to absorb up to $9 \times 0.5 = 4.5$ W of power.

In order to permit the user to connect/disconnect the transfer of energy from the PS to the circuit under test, (generally but not always) there is a button on the front panel of the PS. This button connects/disconnects the load voltage (Fig. 2.12).

2.6   METERS OF THE PS

Each variable channel of a PS has a voltmeter and an ammeter (Fig. 2.13). When the PS supplies the load, the voltmeter and ammeter show the value of load voltage and current. These meters have another function: when you want to set the channel voltage and maximum current, they show the values for you so you can set the PS to values you want.

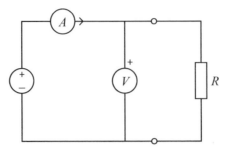

Figure 2.13: Connection of an ammeter and voltmeter to the PS. Output power is the multiplication of the ammeter reading into the voltmeter reading.

Figure 2.14: A DMM could be used to measure the output voltage of the PS accurately.

If you prefer, you can use a DMM in order to measure the voltage/current of load. The DMM permits you to read with higher accuracy. As shown in Fig. 2.14, the meter on the front panel of PS shows 9.0 V while the DMM reads 9.03 V.

2.7   WORKING MODES OF PS'S

Assume that we have two similar and independent DC power sources, as shown in Fig. 2.15.
Voltage of each power source could vary between zero and a maximum value:

$$0 < V1 < V \max$$

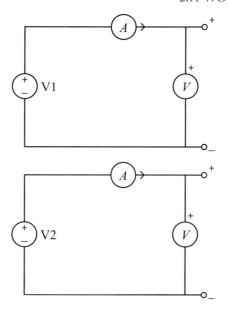

Figure 2.15: Two independent DC power sources.

$$0 < V2 < V \max.$$

The outputs current that each PS could provide is limited between zero and a maximum value as below:

$$0 < I1 < I \max$$

$$0 < I2 < I \max.$$

We could use these two DC power sources in three different ways: Independent, parallel, and series.

2.7.1   INDEPENDENT MODE

In this case, we could supply two different loads (Fig. 2.16). Voltage of each load is between zero and Vmax. The load current could be between zero and I max.

2.7.2   PARALLEL MODE

Assume that we connected the DC voltage sources in parallel, as shown in Fig. 2.17. In this case, the load voltage could be between zero and Vmax. However, this parallel connection could give more current. It could give up to $I \max + I \max = 2I \max.$

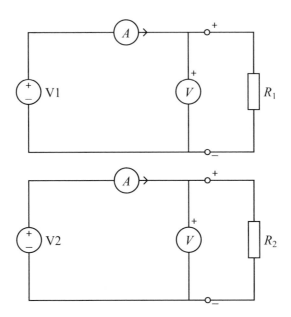

Figure 2.16: Independent operating mode.

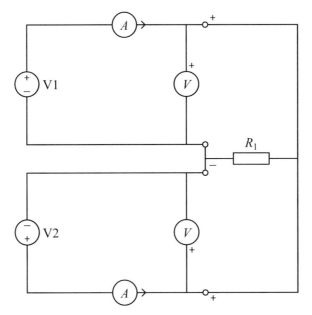

Figure 2.17: Parallel operating mode.

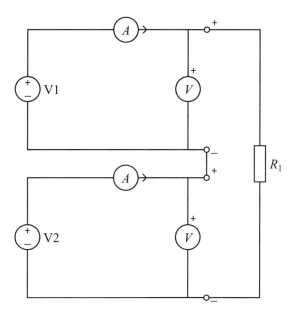

Figure 2.18: Series operating mode.

2.7.3   SERIES MODE

Assume that we connected the DC voltage sources in the series, as shown in Fig. 2.18. In this case, the load voltage could be between zero and V max $+V$ max $= 2V$ max. However, in the series connection, the output current could be between zero and I max.

Let's summarize: when you need more voltage, you need to use the series connection. When you need more current, the you need to use the parallel connection. These connections are used in PSs as we will see in the next sections.

2.8   SAMPLE PS

In this section we will study a sample PS. The model used here is GW INSTEK GPS-3303. This PS is shown in Fig. 2.19.

The outputs of this PS is shown in Fig. 2.20. This model has three channels. The labels behind each channel help you to know whether the channel has a fixed voltage or its voltage is variable. CH 1 and CH 2 have the "0–30 V, 3 A" label, so we deduce that these channels could provide voltages from 0–30 V and the maximum of current they could give us is 3 A (Figs. 2.21 and 2.22). The CH 3 has the "5 V, 3 A" label. So, CH 3 is has a constant voltage of 5 V and could give us up to 3 A of current (Fig. 2.23). The jack with the GND label is connected to the protective earth. The protective earth of the circuit under test is connected to the protective earth of PS.

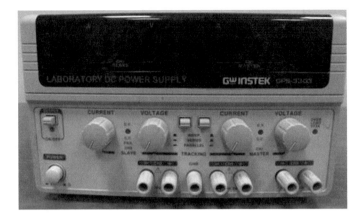

Figure 2.19: GW INSTEK GPS-3303 model PS.

Figure 2.20: Output jacks of GW INSTEK GPS-3303. Protective earth is shown with GND label.

Figure 2.21: Channel 1 of GW INSTEK GPS-3303. According to the label, this channel could provide 0–30 V with maximum current of 3 A.

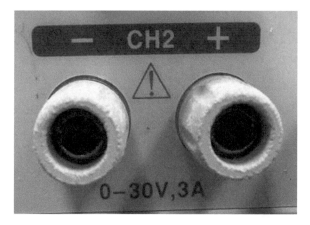

Figure 2.22: Channel 2 of GW INSTEK GPS-3303. According to the label, this channel could provide 0–30 V with maximum current of 3 A.

Figure 2.23: Channel 3 of GW INSTEK GPS-3303. According to the label, this channel could provide fixed 5 V with maximum current of 3 A.

CH 1 and CH 2 are equipped with an ammeter and voltmeters (Fig. 2.24), however CH 3 has no voltage/current indicator. This channel only provides a constant nominal +5 V. You can use a DMM in order to measure its voltage/current when necessary (Fig. 2.25).

The variable channels (CH 1 and CH 2) could be used in three different modes: independent, series, and parallel. The operating mode of the PS is determined using the two buttons shown in Figs. 2.26 and 2.27. When the two buttons are not pressed, then the PS is in the independent mode. It means that you can set the voltage and maximum current of each channel

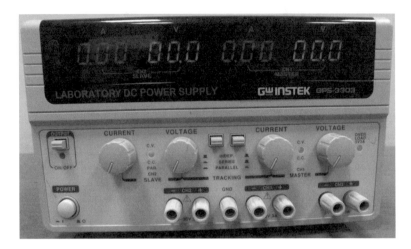

Figure 2.24: CH 1 and CH 2 have an ammeter and voltmeters in order to show the output current and voltage.

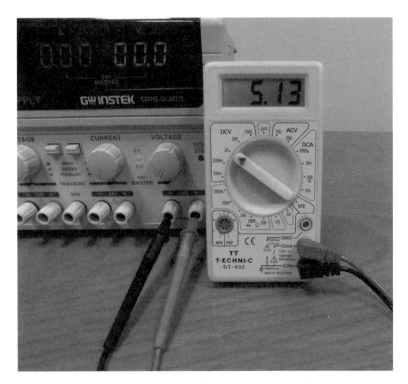

Figure 2.25: Measurement of the channel 3 voltage with a DMM.

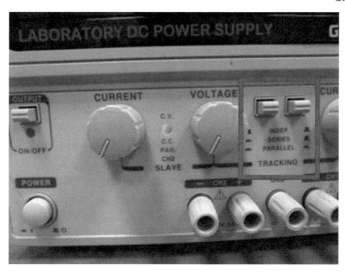

Figure 2.26: Buttons to set the operating mode of the PS.

Figure 2.27: Close up of the mode selection unit of PS.

independently. For instance, you can set the CH 1 for 12 V and 1 A and CH 2 for 7 V and 200 mA.

If the left button is pressed but the right-hand side button is not pressed, then the PS is in the series mode. In this case, you can have a voltage between 0 and 60 V with maximum output current of 3 A. When you are in the series mode, only the CH 1 controls (voltage and current knob) are active. In this case, CH 2 obeys the CH 1 settings. For instance, if you set CH 1 for 18 V, 2.5 A, then the same settings will be applied to CH 2. That is why CH 2 has the SLAVE label (see Fig. 2.24). In this case, the output voltage is the summation of CH 1 and CH 2 voltages, so the output voltage will be $18 + 18 = 36$ V. The maximum output current is the same as channel 1 current, so the maximum current is 2.5 A. In the series mode, the output voltage is taken from the positive output (red jack of CH 1) of CH 1 and negative output of CH 2 (black jack of CH 2).

If both buttons are pressed, then the PS is in the parallel mode. In this case, you can have a voltage between 0 and 30 V with maximum output current of 6 A. When you are in the parallel mode, only the CH 1 controls (voltage and current knob) are active. Like the series case, the CH 2 obeys the CH 1 settings. For instance, if you set CH 1 for 18 V, 2.5 A, then the same settings will be applied to CH 2. In this case, the output voltage is the same as the voltage of CH 1, however the maximum output current could be as high as $2.5 + 2.5 = 5$ A. In the parallel mode, the output voltage is taken from the CH 1 jacks. You can use the CH 2 jacks as well because the positive output of CH 1 is connected to the positive output of CH 2 and the negative output of CH 1 is connected to negative output CH 2. However, using the master channel outputs are better because the control system of PS measures the CH 1. So, when you connect the external circuit to CH 1 output, the load current divides almost equally. However, when you connect the external circuit to CH 2, there is a little bit of unbalance between the current drawn from channels.

So, using these buttons you could have 0–60, 3 A, and 0–30 V, 6 A (Fig. 2.26 and Fig. 2.27).

Assume we want to set the CH 1 for 9 V and 500 mA. In order to do this, you need to rotate the voltage knob of CH 1 until you obtain the desired voltage. As you rotate the voltage knob, the voltmeter of CH 1 measures the output voltage and displays it for you. Before setting the desired output voltage (which is 9 V in this example), set the output voltage to a low level (for instance 2 V). Then short circuit the output of channel (connect the alligator clips of the output cable together) and use the current knob in order to set the maximum current of channel. In our example this maximum is 500 mA or 0.5 A. So, we rotate the current knob until we amprer meter of corresponding channel shows 0.5 A. When we reached that point, then we stop rotation. After setting the maximum current, open the short circuit you made and use the voltage knob in order to set the output voltage to desired value.

Assume that you set the output voltage to 9 V and maximum of 500 mA. In this case, if you connect a load with resistance bigger than $\frac{9}{0.5} = 18$ Ω, i.e., $18 < R < \infty$, then the output voltage will be kept constant since such a values of load require current less than the maximum. So, in this case the PS acts as a constant voltage source (with voltage of 9 V). Generally, the constant voltage mode of operation is shown with an LED indicator with the CV (Constant Voltage) label.

For the $0 < R < 18$ range, the PS acts as a constant current source. So, the load current is constant, however the load voltage is variable. For instance, if you connect a 7 Ω load to the PS, the load voltage will be $7 \times 0.5 = 3.5$ V. Like the constant voltage mode, the constant current mode of operation has an indicator. This indicator has the CC (Constant Current) label.

When the CC indicator is active, you understand that your circuit wants to draw more current, however the protection circuit of the PS limits this current to the value that you determine using the current knob of the channel. One of the most common reasons for this more current demand is a short circuit. So, keep in mind that the CC indicator is an alert for you.

2.9   PARALLEL MODE OPERATION

As said before, the parallel mode of operation permits you to supply loads up to 6 A. For instance, assume that you want to produce 5 V with maximum current of 5.5 A. In order to do this do the following:

1. Use the buttons shown in Fig. 2.26 in order to select the parallel mode of operation.

2. In the parallel mode of operation, output voltage is taken from the positive jack of the master channel and negative jack of the master channel. So, connect the output cable there. You can connect the output cable to the slave channel outputs. However, this is not recommended because when you use the slave outputs, the load current may not be divided equally between the two channels (see Figs. 2.28 and 2.29).

3. In the parallel mode of operation, only the control knobs of the master channel are active. Set the voltage to a small value, i.e., 2 V. Short circuit the wires together in order to set the maximum value of the current. Set the current control of the master channel to half of the desired maximum current. For instance, for a maximum of 5.5 A, set the current knob of the master channel to 2.75 A. Since you are in parallel mode, the output current will be the summation of these currents, i.e., $2.75 + 2.75 = 5.5$ A (Fig. 2.30).

4. After setting the maximum value of current, you can set the desired voltage.

2.10   SERIES MODE

As said before, the series mode of operation permits you to have output voltage up to 60 V. For instance, assume that you want to produce 50 V with a maximum current of 2 A. In order to do this do the following:

1. Use the buttons shown in Fig. 2.26 in order to select the series mode of operation.

2. In the series mode of operation, output voltage is taken from the positive jack of the master channel and negative jack of the slave channel (Fig. 2.31). So, connect the output cable there.

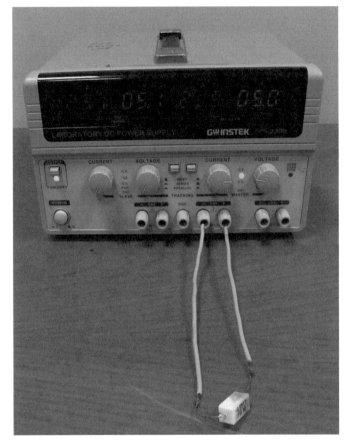

Figure 2.28: When output is taken from the master channel, the load current is divided almost equally between the two channels.

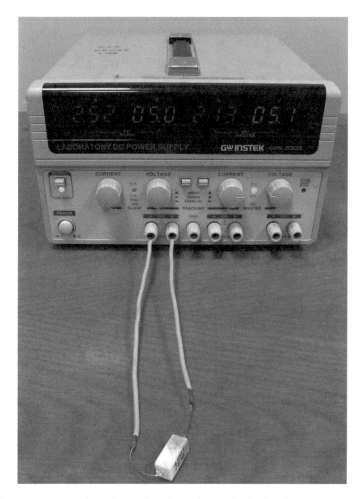

Figure 2.29: When output is taken from the slave channel, the load current is not divided equally between the two channels.

Figure 2.30: The maximum output is set to 2.75 A + 2.75 A = 5.5 A.

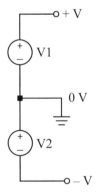

Figure 2.31: You can use the series mode of operation in order to produce the symmetric voltages. V1 is the master channel and V2 is the slave channel.

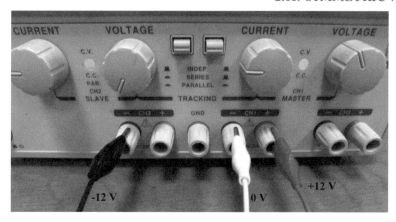

Figure 2.32: The series also could be used to obtain symmetric voltages. 0 V is taken from the negative output of channel 1.

3. In the series mode of operation, only the control knobs of master channel are active. Set the voltage to a small value, i.e., 2 V. Short circuit the wires together in order to set the maximum value of the current. Here, the maximum value of current is given as 2 A. So, increase the current knob of master channel until you see 2 A on the front panel current display.

4. After setting the maximum value of the current, you can set the desired voltage. Set the master channel to half the desired output voltage. For instance, in order to obtain 50 V, set the master channel to 25 V. Since you are in series mode, the output voltage will be the summation of channels 1 and 2. So, $25 + 25 = 50$ V.

2.11 SYMMETRIC VOLTAGES

Some circuits (especially opamp circuits) require a symmetric voltage, for instance $+12$ V, -12 V. You can use the series mode of operation in order to produce such a symmetric voltage (see Fig. 2.31). The positive jack of the master channel gives the positive voltage that you want. The negative jack of the slave channel gives the negative voltage that you want. The positive jack of the slave channel or negative jack of the master channel gives you the 0 V, i.e., ground (see Figs. 2.32 and 2.33).

2.12 FURTHER READING

[1] John Lenk, *Simplified Design of Linear Power Supplies*, Newnes, 1994.

[2] Paul Lee, *Power Supplies Explained*, Radio Society of Great Britain, 2018.

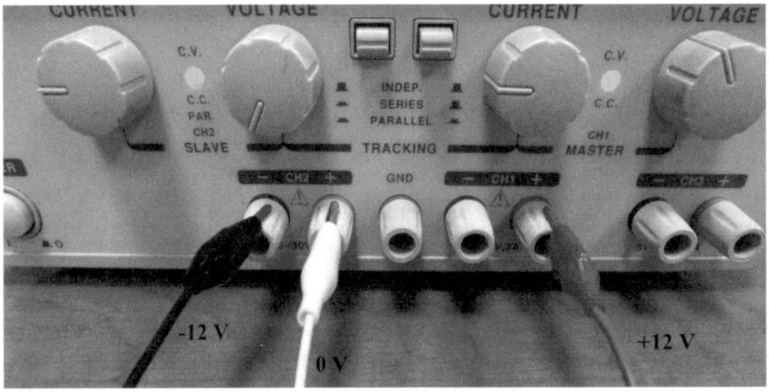

Figure 2.33: 0 V can be taken from the positive output of channel 2.

CHAPTER 3

Function Generator

3.1 INTRODUCTION

In order to test an electronic circuit, you need test signals. For instance, assume you designed an amplifier and you want to measure its voltage gain. In order to measure the voltage gain, you need to feed the input of amplifier with a signal (for instance, a sinusoidal signal) with known amplitude and measure the amplitude of output. Division of output amplitude to the input amplitude gives the voltage gain of the amplifier under test. So, without a test signal, you can't measure the voltage gain of the amplifier.

As another example, assume that you designed a (sequential) digital circuit. When you want to test it, you need to apply a square wave (clock pulse) to the ICs that you used. So, you need a tool in order to produce the required square wave pulse for you.

These two simple examples show the need for a device that is capable of producing different signals. Such a device is called a Function Generator (FG). An FG (sometime called a signal generator) is used to generate different types of electrical waveforms over a wide range of frequencies. Some of the most common waveforms produced by the FG are the sine wave, square wave, triangular wave, and saw tooth shapes (Fig. 3.1).

The FGs are divided into two groups: analog FGs and Direct Digital Synthesis (DDS) FGs. As the name suggests, the analog FGs use the analog circuits in order to produce the output waveform. DDS FGs uses digital circuits (i.e., a microprocessor) in order to produce the output waveforms. Accuracy of the DDS signal generator are better in comparison to analog signal generators. Besides the standard waveforms (i.e., sinusoidal, square, triangular, and saw tooth), some DDS FGs are able to produce arbitrary waveforms. These types of FGs are called Arbitrary Waveform Generators (AWGs). They have software which permits you to draw the waveform that you want. After drawing the waveform in the software environment, the hardware of AWG produces the waveform for you (Fig. 3.2).

3.2 FG'S OUTPUT CABLE

The FGs require a cable in order to transfer the signal to the circuit under test. Such a cable has a Bayonet Neill–Concelman (BNC) connector on one end and alligator clips on the other end. The BNC connector is used to connect the cable to the FG and the alligator clips are connected to the circuit under test (see Figs. 3.3 and 3.4).

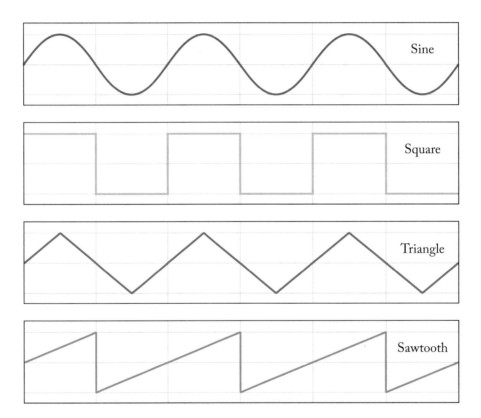

Figure 3.1: Typical test waveforms.

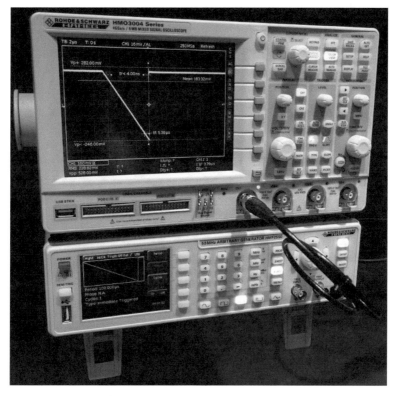

Figure 3.2: HAMEG HMF 2550 is an example of AWG.

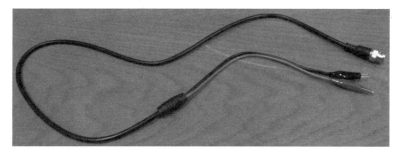

Figure 3.3: Output cable of FG.

Figure 3.4: Output cables of FG have a BNC connector on one end and alligator clips on the other end.

The output of FG could not supply huge amounts of current. A typical FG could give up to 100–150 mA of current. Generally, the output resistance of FG is 50 Ω (Fig. 3.5).

You can use a resistor (a resistor with value of 33–50 Ω is suggested) in order to measure the output resistance of FG. In order to do this, do the following.

1. Measure the open circuit output voltage (Voc) of the FG (Fig. 3.6). You can use an oscilloscope or DMM for this purpose. If you are using an oscilloscope, you can measure the peak value of waveform. If you are using a DMM, you can measure the RMS value of output voltage.

2. Connect the load resistor to the output of the FG and measure the voltage across the load resistor (Vout) (Fig. 3.7). Measurement must be done with the same device as in step 1. For instance, if you used an oscilloscope and measured the amplitude of signal in step 1, then you need to use an oscilloscope in this step as well and measure the amplitude of resistor voltage.

Figure 3.5: Generally, the output resistance of FGs is 50 Ω.

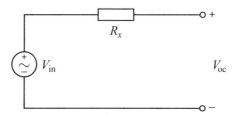

Figure 3.6: Measurement of open circuit voltage. Rx shows the internal resistance of FG.

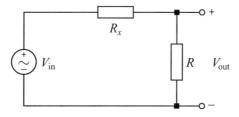

Figure 3.7: Measurement of load voltage. The value of load resistor is known.

3. Solve the following equation in order to find the unknown value (R_x):

$$\frac{R}{R_x + R} V_{oc} = V_{out} \implies R_x = \frac{R \times V_{oc}}{V_{out}} - R.$$

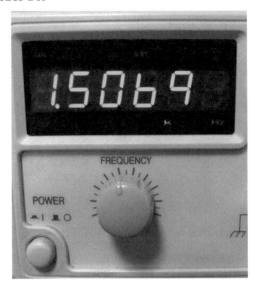

Figure 3.8: The output frequency of the FG is displayed.

3.3 OUTPUT FREQUENCY DISPLAY

Some FGs have a display on the front panel which shows the frequency of the output (Fig. 3.8). This helps the user to set the output frequency easily. If the FG doesn't have a frequency display, then you need to use an oscilloscope (or a DMM with frequency counting capability) in order to set the frequency.

3.4 −20 dB BUTTON

Assume that you want to measure the voltage gain of an amplifier. In this case you need to supply the input of the amplifier with a small signal with amplitude in the range of few 10 mV. Otherwise, the output of your amplifier will be saturated. Assume that you need a sinusoidal signal with peak of 20 mV in order to test the amplifier. Simply use the amplitude knob of the FG and produce a signal with amplitude of 200 mV (you need an oscilloscope in order to set the amplitude to 200 mV). Then, activate the −20 dB button (see Figs. 3.9 and 3.10) in order to decrease the amplitude of output by 10.

You may ask "Why we don't set the output for 20 mV directly?" The answer is: setting the output for small values is difficult. The user uses his/her hand in order to rotate the amplitude knob of FG and a small movement in that knob may change the output with few tenth of milli volts. So, direct setting for small values is difficult. Instead of directly setting the output to the value that we want, we set it to a value that is ten times bigger and press the −20 dB button. If

Figure 3.9: When you pull this knob, the output is decreased by a factor of 10.

Figure 3.10: When you push this button, the output is attenuated by a factor of 10.

Figure 3.11: The "output pulse" output. As the name suggests, this output only gives pulses.

you are using a digital FG, then you can enter the desired amplitude value that you want directly with the aid of keypad in the front panel of the device.

3.5   OUTPUT PULSE

Generally, the analog FG has a second output which is called "Output Pulse" (Fig. 3.11). This output only produces pulse waveforms. For instance, if you need a clock pulse for a digital circuit, you can use this output.

3.6   OFFSET AND DUTY RATIO KNOBS

In this section we study the function of offset and duty ratio knobs.

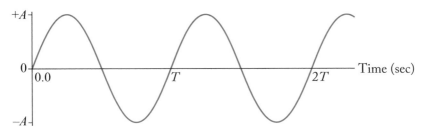

Figure 3.12: Plot of $A \sin \left(\frac{2\pi}{T} t\right)$.

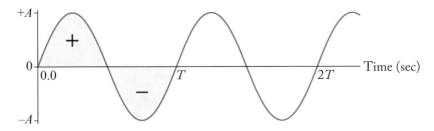

Figure 3.13: The positive half cycle is equal to the negative half cycle.

3.6.1   OFFSET

DC offset (sometimes called DC bias, DC component, or average value) is the mean amplitude of a waveform. For instance, take a look at the sinusoidal waveform that is shown in Fig. 3.12.

Consider one period of this waveform. Half of the period is positive and half of the period is negative. Since the area of positive half cycle and negative half cycles are equal, the summation of these two components is zero (Fig. 3.13).

Let's say what we said in mathematics language. In mathematics, the average value of a periodic signal $f(t) = f(t + T)$ is defined as:

$$\frac{1}{T} \int_{t_0}^{t_0+T} f(\tau)d\tau.$$

T shows the period of signal. The average value of a sinusoidal signal is:

$$\frac{1}{T} \int_{0}^{T} A \sin \left(\frac{2\pi}{T} \tau\right) d\tau = 0.$$

Now consider that we added a constant term to the aforementioned sinusoidal signal (Fig. 3.14).

In this case, the average value of $V_{dc} + A \sin(\frac{2\pi}{T} t)$ is not zero because the positive area is quite bigger than the negative area (Fig. 3.15).

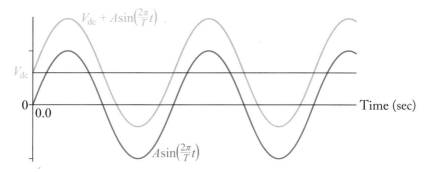

Figure 3.14: Graph of $A \sin(\frac{2\pi}{T}t)$ and $V_{dc} + A \sin(\frac{2\pi}{T}t)$. V_{dc} is a positive number.

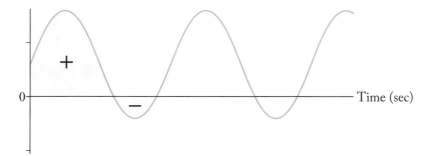

Figure 3.15: DC offset is positive.

Using the mathematical definition of average value, we obtain:

$$\frac{1}{T} \int_0^T V_{dc} + A \sin\left(\frac{2\pi}{T}\tau\right) d\tau = V_{dc}.$$

So, when you add a constant DC term to your periodic waveform, this constant DC term is called offset value. When the positive half cycle of periodic signal has the same area as the negative half cycle (like Fig. 3.13), the offset (or average value) is zero. If the area above the time axis is bigger than the area under the time axis, then the summation will be a positive value. In this case, the offset is positive. If the area of part of signal which lays down the time axis is bigger than the part above the time axis, then the net result will be a negative number. In this case the, the offset is negative (see Fig. 3.16).

Using the offset knob of FG, you can add DC value to your periodic signal. Figure 3.17 shows a simple equivalent circuit for the output of the FG. The value of Vdc in Fig. 3.17 is controlled with the aid of the offset knob of the FG. Amplitude of AC source (Vac) is controlled with the aid of the amplitude knob of the FG. AC source could be a sinusoidal, pulse, triangular, or saw tooth waveforms.

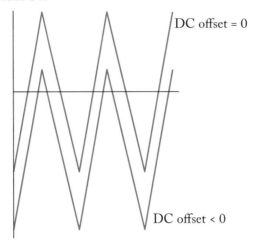

Figure 3.16: DC offset is negative.

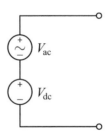

Figure 3.17: A waveform could be imagined as a series connection of an AC source (with zero average) and a DC source.

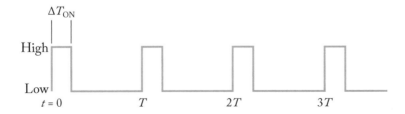

Figure 3.18: A typical pulse.

3.6.2   DUTY RATIO

Consider a pulse like the one shown in Fig. 3.18.

High
Low
$t = 0$ $T = 5\ \mu s$ $T = 25\ \mu s$

Figure 3.19: Duty ratio is $D = \frac{5\ \mu s}{25\ \mu s} = 0.2$ or 20%.

Figure 3.20: DUTY knob of FG.

The frequency of this pulse is $f = \frac{1}{T}$. The duty ratio (D) is defined as:

$$D = \frac{\Delta T_{ON}}{T}.$$

Sometimes it is defined in percentage:

$$D = \frac{\Delta T_{ON}}{T} \times 100\%.$$

For instance, for the pulse shown in Fig. 3.19, the duty ratio is: $D = \frac{5\ \mu s}{25\ \mu s} = 0.2$ or 20%.

Generally, FGs have a control called "Duty" (Fig. 3.20). Using this control, you can change the duty ratio of pulses. Using the duty control, you can also convert a symmetrical triangular waveform (Fig. 3.21) to a non-symmetrical waveform (Fig. 3.22) as well.

3.7   VOLTAGE CONTROL FREQUENCY (VCF) INPUT

Using the VCF input of FG, you can control the output frequency with a control voltage (Fig. 3.23).

3.8   SAMPLE FG

In this section, GW INSTEK GFG-8015G FG is studied as an example (Fig. 3.24).

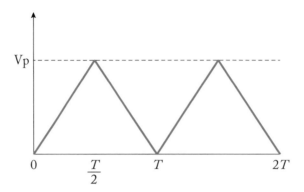

Figure 3.21: **Symmetrical waveform.**

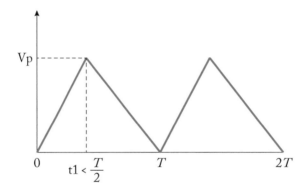

Figure 3.22: **Asymmetrical waveform.**

Figure 3.23: **VCF input of FG.**

Figure 3.24: GW INSTEK GFG-8015G model analog FG.

Figure 3.25: Power button.

3.9   GENERATION OF DESIRED WAVEFORM

Assume that we want to produce a sinusoidal waveform with GW INSTEK GFG-8015G FG. In order to produce a desired waveform, do the following.

1. Press the "PWR" button in order to turn on the FG (Fig. 3.25).

2. Connect the output cable to "OUTPUT 50 Ω" (Fig. 3.26).

3. Use the "FUNCTION" buttons in order to select the type of waveform which you want (Fig. 3.27).

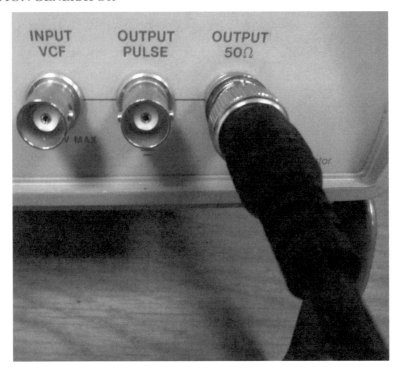

Figure 3.26: Output cable is connected to the 50 Ω output.

Figure 3.27: Output waveform selector.

4. The output frequency is:

$$f = R \times P.$$

R is the pressed button of "Range Hz" section (see Fig. 3.28). *P* is the value of frequency knob (see Fig. 3.29). For instance, if you press the 1 K button from the "Range Hz" section and set the frequency knob to 1, then the output frequency will be:

$$f = R \times P = 1 \text{ K} \times 1 = 1 \text{ kHz}.$$

Figure 3.28: Range Hz section.

Figure 3.29: Frequency knob.

You can use a DMM with frequency counting capability or an oscilloscope in order to measure the output frequency of the FG (Fig. 3.30).

5. Connect the output of FG to an oscilloscope (Fig. 3.31). Use the FG's "AMPL" control (see Fig. 3.32) in order to set the desired output peak. The obtained waveform is shown in Fig. 3.33.

3.10   OFFSET KNOB

Use the OFFSET knob (Fig. 3.34) of the FG in order to add an offset (DC value) to the output.

In this FG, you need to pull out the OFFSET knob in order to activate it (Fig. 3.35). After the OFFSET knob is pulled out, connect a DMM to the output of the signal generator (red alligator clips of the output cable are connected to the red probe of the DMM; see Fig. 3.36).

Figure 3.30: Output frequency is 0.986 kHz or 986 Hz.

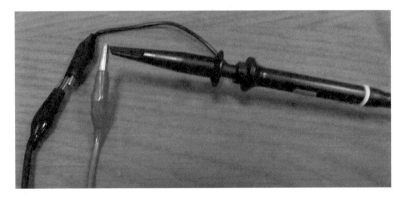

Figure 3.31: An oscilloscope can be used to measure the output frequency and amplitude.

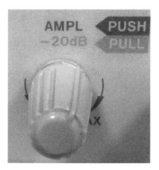

Figure 3.32: Amplitude knob. This knob controls the amplitude of the output waveform.

Figure 3.33: The output waveform.

Figure 3.34: Offset knob.

Figure 3.35: Pull out the offset knob in order to activate it.

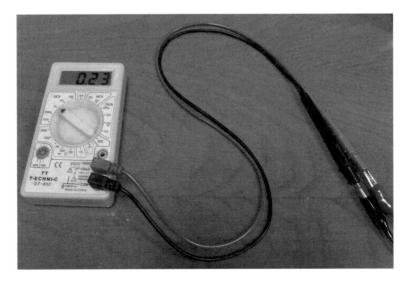

Figure 3.36: Measurement of the output DC value. Note that the DMM must be in the DC voltage measurement mode.

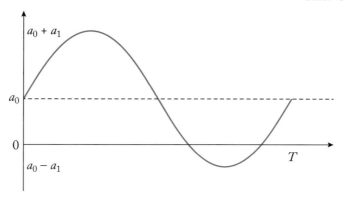

Figure 3.37: Graph of $f(t) = a_0 + a_1 \sin(\omega t)$. Maximum and minimum of signal are $a_0 + a_1$ and $a_0 - a_1$, respectively. The offset of this signal is a_0.

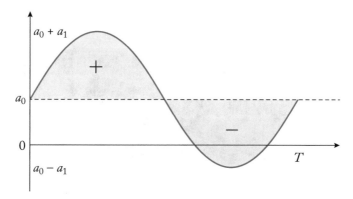

Figure 3.38: The area above and below the $y = a_0$ line are the same. So, the offset or DC term is a_0.

Set the DMM in the DC voltage measurement mode. Now, rotate the OFFSET knob until you see the desired offset value on the DMM display.

You can set the desired offset value with an oscilloscope as well. A graphical method is introduced for this purpose: when the integral symmetry axis of the signal intersects the vertical axis at $y = a_0$, then the DC term or offset of the signal will be a_0 (see Fig. 3.37). By integral symmetry axis we mean a horizontal line which the area of the part of the signal which is above the line equals to the area of the part of the signal which is under the line (see Fig. 3.38).

For instance, the blue signal in Fig. 3.39 has zero offset while the green signal in Fig. 3.39 has 3 V offset.

Note: For GW INSTEK GFG-8015G, the output voltage could be between -13 and $+13$ V. If you start to increase the offset, the output will be clipped (Fig. 3.40).

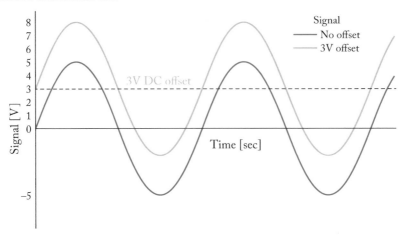

Figure 3.39: The equation of blue signal is $f(t) = 5\sin(\omega t)$ and the equation of green signal is $g(t) = 3 + 5\sin(\omega t)$.

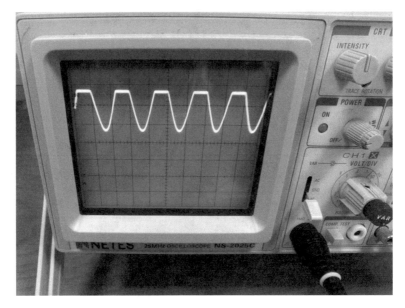

Figure 3.40: The output clipped.

3.11   DUTY KNOB

Using the DUTY control (see Fig. 3.41), you can produce asymmetric waveforms. In order to activate the DUTY control, turn the DUTY knob to right. Connect the output of FG to an oscilloscope (see Fig. 3.42). Now you can see the changes as you rotate the DUTY knob. Using

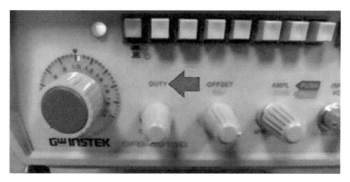

Figure 3.41: DUTY knob.

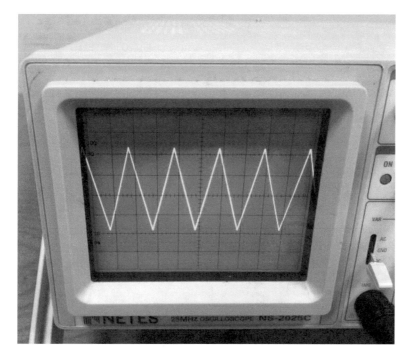

Figure 3.42: Symmetric waveform.

the DUTY control you can change a waveform like Fig. 3.42 into Fig. 3.43 or Fig. 3.44. As another example, you can change the symmetrical square waveform Fig. 3.45 into Fig. 3.46. When you are rotating the DUTY knob, the frequency of the signal is not changing. The only thing that is affected is the symmetry of the signal.

In order to deactivate the DUTY control and obtain a symmetric waveform again, rotate the DUTY knob to the far-left point.

Figure 3.43: Asymmetric waveform.

Figure 3.44: Asymmetric waveform.

Figure 3.45: Symmetric pulse.

Figure 3.46: Asymmetric pulse.

Figure 3.47: The ouput cable is connected to the pulse output of the FG.

Figure 3.48: The ouput pulse has a high value of 5 V and low value of 0 V.

3.12   "OUTPUT PULSE" OUTPUT

Using the "OUTPUT PULSE" output of the FG (Fig. 3.47) you can generate pulses with a high level of +5 V and a low level of 0 V (Fig. 3.48). This type of pulse is suitable for (TTL compatible) digital circuits. The amplitude of "OUTPUT PULSE" is not under control, i.e., the +5 V and 0 V are not changeable, however you can control the frequency and duty ratio of the output waveform (Fig. 3.49). Note that you cannot take sinusoidal, triangular, or saw tooth signals from this output. You can only take pulse waveforms from this output.

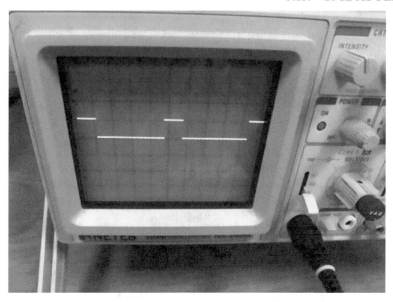

Figure 3.49: The duty ratio of output pulse could be changed.

Figure 3.50: When you pull out the amplitude knob, the −20 dB activates.

3.13   −20 dB ATTENUATION

Using the −20 dB button, you can multiply the current output with 0.1. In order to activate the −20 dB attenuation, pull out the amplitude (AMPL) knob (see Fig. 3.50). For instance, the waveform shown in Fig. 3.51 has an amplitude of 2 V. After pulling out the amplitude knob, the output changes from 2 V into 0.2 V (see Fig. 3.52).

   As said before, the −20 dB attenuation permits you to set the required small amplitudes more easily. Direct setting the small amplitudes, i.e., amplitudes in the few 10 mV range, may be difficult. So, use the −20 dB when you need to produce a small output. To do this, set the output for a value that is 10 times bigger and after that activate the −20 dB button. (For instance, for

Figure 3.51: Before pulling out the AMPL knob, the output peak is 2 V.

Figure 3.52: After pulling out the AMPL knob, the output peak is 0.2 V.

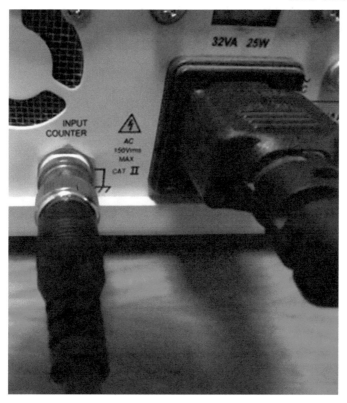

Figure 3.53: Input counter jack.

a 20 mv output, you need to set the output to 200 mv). In some FGs, you may even see bigger attenuations, for example −40 dB attenuation which equals to decreasing the output by factor of 100.

3.14   COUNTER MODE

Some FGs (e.g., GW INSTEK GFG-8216A) could be used as frequency meter, i.e., for measuring the frequency of a signal. Note that in the counter mode, the FG doesn't produce any signal; instead it measures the frequency of input signal.

In order to use the frequency counter mode, you need to find a jack with the "INPUT COUNTER" label (Fig. 3.53). This jack generally is placed at the back of FG, behind the power cord.

3.15   FURTHER READING

[1] David Bell, *Solid State Pulse Circuits*, Prentice Hall, 1991.

CHAPTER 4

Oscilloscope

4.1 INTRODUCTION

The oscilloscope is most likely the most important measurement device. Oscilloscopes permit you to see the voltage waveforms. The oscilloscopes could be divided into two groups: analog oscilloscopes and digital oscilloscopes. Examples of an analog oscilloscope and digital oscilloscope are shown in Figs. 4.1 and 4.2, respectively.

Analog oscilloscopes use a Cathode Ray Tube (CRT) in order to show the waveforms. A CRT sample is shown in Fig. 4.3. A CRT is a glass envelope which is deep, heavy, and fragile. The interior is evacuated to approximately 0.01 Pascal's to 133 nano Pascal's to facilitate the free flight of electrons from the gun(s) to the tube's face without scattering due to collisions with air molecules. The face is typically made of thick lead glass or special barium-strontium glass to be shatter-resistant and to block most X-ray emissions. CRTs make up most of the weight of an analog oscilloscope.

The structure of CRT is shown in Fig. 4.4. The input waveform to the oscilloscope with the aid of deflecting coils deflects the electron beam coming from the heated cathode and after the electrons hit the fluorescent screen, an image appears on the screen. Analog oscilloscopes are heavier than digital oscilloscopes.

Digital oscilloscopes don't use CRT. The digital oscilloscopes use LCD screens. The digital oscilloscopes use Analog to Digital Converter (ADC) in order to sample the input waveform. These samples will be used in order to form the image on the screen (Fig. 4.5).

The digital oscilloscopes are more flexible than the analog ones. They could do a lot of measurements automatically. For instance, they could measure RMS or average value of a complex signal, measurement of duty ratio of pulses, measurement of peak-to-peak voltage measurement, measurement of settling time of waveforms, etc.

These days, the price of a digital oscilloscope is quite affordable. So, they took the place of analog oscilloscopes.

4.2 CHANNELS

Channels are the input path for external signals to enter the oscilloscope. For instance, a 2-channel oscilloscope has two inputs for external signals to enter the oscilloscope. In other words, a 2-channel oscilloscope is an oscilloscope which is able to show 2 separate waveforms simul-

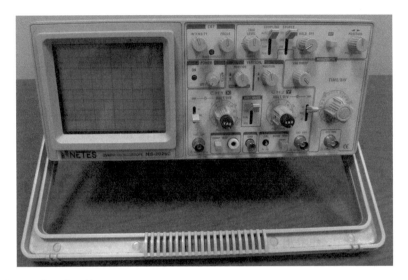

Figure 4.1: An analog oscilloscope.

Figure 4.2: A digital oscilloscope.

Figure 4.3: CRT of an analog oscilloscope.

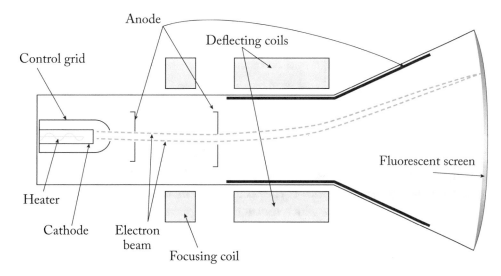

Figure 4.4: Structure of CRT.

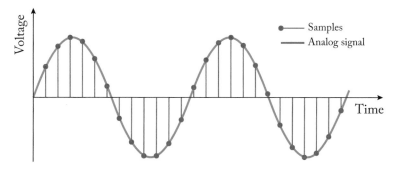

Figure 4.5: Sampling an analog signal.

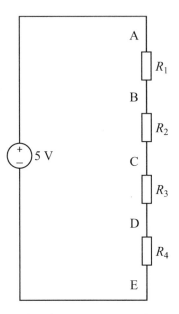

Figure 4.6: Simple voltage divider. All the resistors are the same. $V_{AB} = V_{BC} = V_{CD} = V_{DE} = \frac{5}{4}$ V $= 1.25$ V.

taneously. Input channels of oscilloscopes are shown with CH 1, CH 2, …. Note that as the number of channels increases, the price of the oscilloscope increases as well.

Each channel requires a probe which transfers the signal from the circuit under test to the oscilloscope. Note that the ground of all the channels are connected to each other. This could be understood with the aid of the simple circuit shown in Fig. 4.6. All the resistors are the same, for instance R1 = R2 = R3 = R4 = 1 kΩ and the input voltage is 5 V.

Assume that someone connected the tip of the probe of channel 1 to point A, the ground of probe of channel 1 to point B, tip of probe of channel 2 to point D, and the ground of probe of channel 2 to point E. He/she expects the scope to show the voltage across the resistor R1 on channel 1 and voltage across resistor R4 on channel 2. The voltage drop across R1 is 1.25 V. The voltage drop across R4 is 1.25 V as well. However, the oscilloscope shows +5 V for channel 1 and 0 V for channel 2. Why?

The reason is that when you connect the ground of channel 1 to point B and the ground of channel 2 to E, you connect them together (see Figs. 4.7 and 4.8). So, channel 1 is connected to 5 V and channel 2 will be 0 V.

4.3   OSCILLOSCOPE PROBES

Oscilloscope probes transfer the signal from the circuit under test to the oscilloscope. A typical probe is shown in Fig. 4.9. The end that is connected to the oscilloscope has the BNC connector.

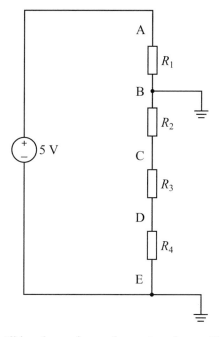

Figure 4.7: Points B and E will be shorted together using the oscilloscope.

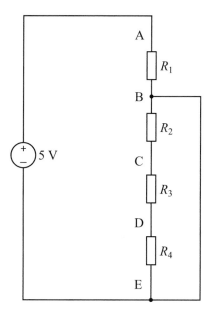

Figure 4.8: Equivalent circuit of Fig. 4.7. $V_A = 5$ V, $V_B = V_C = V_D = V_E = 0$ V.

Figure 4.9: An oscilloscope probe.

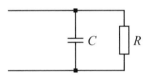

Figure 4.10: Equivalent circuit of oscilloscope input.

The end which connects to the circuit under test has a tip and an alligator clips. The voltage that you see on the oscilloscope is the voltage difference between the tip of probe and the alligator clips, i.e., $V_{tip} - V_{alligator\ clip}$. So, the tip of the probe is like the red probe of DMM and alligator clip is like the black probe of DMM. Usually, the alligator clip is connected to the ground of the circuit under test.

Equivalent front input of an oscilloscope is shown in Fig. 4.10.

The values of the resistor and capacitor are printed on the front panel of the oscilloscopes. The value of R is generally 1 MΩ. The typical value of C is a few pico Farads (Figs. 4.11 and 4.12).

Equivalent circuit of a typical probe is shown in Fig. 4.13. The cable is a coaxial cable which protects the signal from environment noises.

As shown in Fig. 4.13, the probe has a resistor. Assume that you connected the probe to the oscilloscope (see Fig. 4.14). So, a resistive voltage divider is formed. The amount of voltage that the input of oscilloscope sees is $\frac{1M}{1M + R_{probe}} V_{in}$.

Typical oscilloscope probes have a small switch on them. One side of the switch has x1 label and the other side has the x10 label (see Fig. 4.15).

Figure 4.11: Values of R and C for input of NETES NS-2025C analog oscilloscope. R = 1 MΩ and C = 25 pF.

Figure 4.12: Values of R and C for input of GW INSTEK GDS-1022 digital oscilloscope. R = 1 MΩ and C = 15 pF.

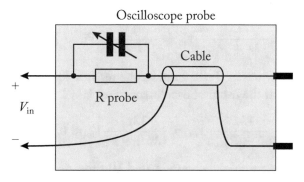

Figure 4.13: Equivalent circuit of oscilloscope probe.

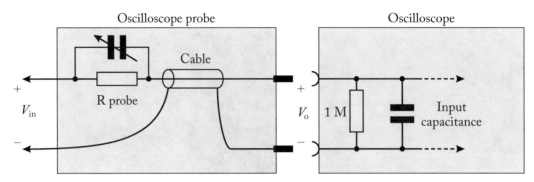

Figure 4.14: Input of oscilloscope makes a voltage divider with the probe resistance.

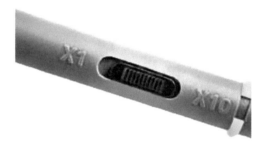

Figure 4.15: Oscilloscope probes usually have a switch on them.

When you put the switch in the x1 mode, the resistor Rprobe ≈ 250 Ω. Figures 4.16 and 4.17 show the measurement of Rprobe for typical probe in the x1 mode.

The voltage that the input of oscilloscope sees is

$$\frac{1M}{1M + R_{probe}} V_{in} = \frac{1M}{1M + 250} V_{in} \approx V_{in}.$$

When you put the switch in the x10 mode, the resistor Rprobe ≈ 9 MΩ (see Fig. 4.18). In this case, the voltage that the input of oscilloscope sees is

$$\frac{1M}{1M + R_{probe}} V_{in} = \frac{1M}{1M + 9M} V_{in} \approx 0.1V_{in}.$$

So, in this case you apply a signal which attenuated 10 times to the input of oscilloscope. For instance, assume that you put the switch in the x10 and you see a sinusoidal signal with peak of 1 V on the screen. In this case, the original signal which is in contact with the probe, has the magnitude which is 10 times bigger, i.e., a signal with peak of 10 V.

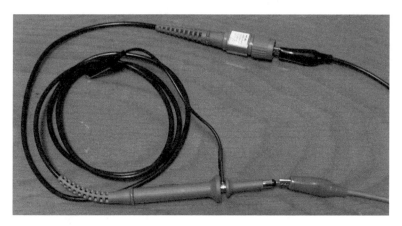

Figure 4.16: Measurement of probe resistor (Rprobe).

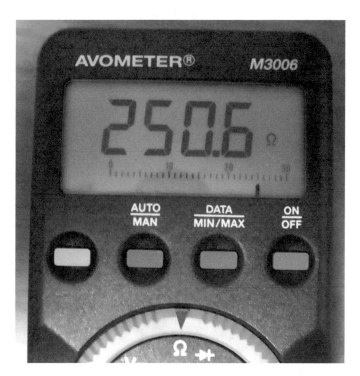

Figure 4.17: Measurement of probe resistor (Rprobe) in the x1 case. Rprobe = 250 Ω.

Figure 4.18: Measurement of probe resistor (Rprobe) in the x10 case. Rprobe = 9 MΩ.

Figure 4.19: Probes have a small variable capacitor for compensation purposes. This is the variable capacitor of Fig. 4.13.

The probe has a variable capacitor as well (see Fig. 4.19). This capacitor is for compensation purposes. Since the input of the oscilloscope has some parasitic capacitance, addition of a small capacitor in parallel with the probe resistance permits better transfer of signal from the external source to the oscilloscope.

Note 1: Each probe has a range of allowed voltages and frequencies which is printed on it. For instance, the probe shown in Fig. 4.20 could measure voltages up to 200 Vpeak in the x1 state and 600 Vpeak in the x10 state safely. Never exceed these ranges.

According to Fig. 4.20, the probe can measure frequencies from DC-60 MHz in the x10 state and it can measure frequencies in the DC-6 MHz range in the x1 mode.

Figure 4.20: Maximum measurable frequency voltage is written on the probe label.

Note 2:   Always ensure that the voltage which enters the oscilloscope is in the allowed range, otherwise the oscilloscope may be damaged. Maximum of allowed voltages are printed behind the input channels of the oscilloscope (see Figs. 4.21 and 4.22). Be conservative and always attenuate the big voltages before applying it to the oscilloscope.

Note 3:   If you pull the front cover of the probe, the tip of probe comes out (see Fig. 4.23). This form of probe is suitable for working with PCBs (see Fig. 4.24). In this case, the tip of probe (which transfers the signal to the oscilloscope) is under your control and you can contact it to where you want.

Note 4:   Attenuation factor of probes is not limited to 1 and 10. There are probes which attenuates the input by factor of 100 or even 1000.

4.4   CALIBRATION POINT

In order to ensure that your scope measures correctly, you need to test your oscilloscope by applying a known test signal and see whether or not the result is correct.

Figure 4.21: The NETES NS-2025C analog oscilloscope could accept voltage with peak values of less than 400 V.

Figure 4.22: The GW INSTEK GDS-1022 digital oscilloscope could accept voltage with peak values of less than 300 V.

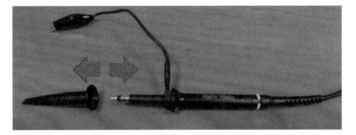

Figure 4.23: Taking out the probe cap.

All the scopes have a 1 kHz square wave generator (amplitude of this square wave changes from one producer to another producer but 2 Vpp is very common value). The output of this generator is accessible from the front panel of the oscilloscope (see Fig. 4.25). It usually has the "CAL" label. CAL stands for calibration.

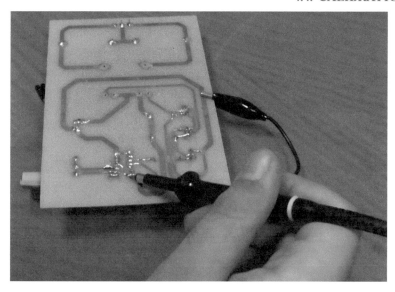

Figure 4.24: A probe without cap could be connected easily to any point of PCB.

Figure 4.25: Calibration point of NETES NS-2025C model oscilloscope.

Figure 4.26: Calibration point of GW INSTEK GDS-1022 model oscilloscope.

It is a good idea to connect the probes to this generator and ensure that what you see on the screen is a square wave with frequency of 1 kHz and amplitude printed on the front panel.

If you see something like the ones shown in Fig. 4.27, then you need to set your probe. In order to set the probe, use a special screw driver which comes with the probe to rotate the variable capacitor on the probe (remember that oscilloscope probes have a variable capacitor, see Fig. 4.28). Change the value of capacitor until you see a waveform like the one shown in Fig. 4.29.

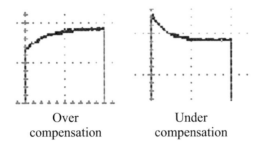

Over
compensation

Under
compensation

Figure 4.27: If you see such waveforms on the screen, then your probe is not set.

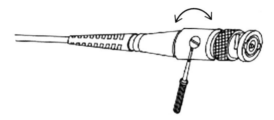

Figure 4.28: Use the small screw behind the BNC connector in order to set the probe.

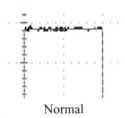

Normal

Figure 4.29: Stop turning the screw when you see such a waveform on the screen.

4.5   MEASUREMENT OF CURRENT WAVEFORMS

An oscilloscope is a device used to see voltage waveforms. An oscilloscope can't show current waveforms. In order to see a current waveform, you need to convert it into a voltage waveform. The simplest way to convert a current signal into a voltage signal is to pass it through a small resistor. When the current passes through the resistor, it generates a voltage drop across the resistor. You can apply this voltage drop to the oscilloscope. According to Ohm's law, the shown waveform on the screen is the scaled version of the current waveform. Using a current probe is another solution to see the current waveform. A current probe is shown in Fig. 4.30.

Figure 4.30: A current probe.

Figure 4.31: Bandwidth of this oscilloscope is 25 MHz.

Figure 4.32: Bandwidth of this oscilloscope is 25 MHz.

4.6   BAND WIDTH OF AN OSCILLOSCOPE

The front input amplifier of an oscilloscope could be considered as a simple low-pass filter with frequency response of

$$H\left(s\right) = \frac{1}{1 + \dfrac{s}{\omega_C}} = \frac{\omega_C}{\omega_C + s}.$$

The ω_C is called the bandwidth of the oscilloscope and refers to the frequency at which the amplitude of the observed signal drops by -3 dB (or drops to 70.7% of its actual value). Value of ω_C is printed on the front panel of oscilloscopes (see Figs. 4.31 and 4.32). The price of an oscilloscope increases as the bandwidth of the oscilloscope increases.

4.7   TIME AND VOLTAGE MEASUREMENTS

In any oscilloscope there is a screen which shows the waveform. This screen usually has 8 vertical units and 10 horizontal units, and each unit is about 1 cm (Figs. 4.33 and 4.34).

All the oscilloscopes have two important controls: vertical controls and horizontal controls. You have as many vertical controls as the number of channels the oscilloscope have. For

Figure 4.33: Screen of an analog oscilloscope. The oscilloscope is not powered on. The axis is printed on the screen.

Figure 4.34: Screen of a digital oscilloscope. The oscilloscope is not powered on. The axis appear after turning on the device.

Figure 4.35: VOLTS/DIV control for channel 1 of NETES NS-2025C model analog oscilloscope. The red button on the VOLTS/DIV control of analog oscilloscopes must be always turned all the way clockwise, otherwise your measurement will be worthless.

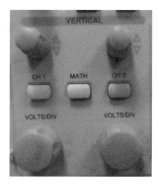

Figure 4.36: VOLT/DIV control of GW INSTEK GDS-1022 model digital oscilloscope.

instance, a 2 channel oscilloscope have 2 vertical controls and a 4 channel oscilloscope have 4 vertical control. Each vertical control belongs to a channel. However, there is only one horizontal control in the oscilloscope.

The vertical controls are called VOLTS/DIV (DIV is taken form the word division). They determine the value of one unit of vertical axis y axis). For instance, when you set the VOLTS/DIV of channel 1 to 5 and channel 2 to 1, then each unit of y axis is 5 V for channel 1 and 1 V for channel 2. The VOLTS/DIV of each channel is under the control of the user. VOLTS/DIV control of an analog oscilloscope and a digital oscilloscope are shown in Figs. 4.35 and 4.36, respectively.

The horizontal control is called TIME/DIV. They determine the value of one unit of horizontal axis (x axis). For instance, when you set the TIME/DIV to 0.2 ms, then each unit of x axis is 0.2 ms (Fig. 4.37). TIME/DIV control of an analog oscilloscope and a digital oscilloscope are shown in Figs. 4.38, 4.39, and 4.40, respectively.

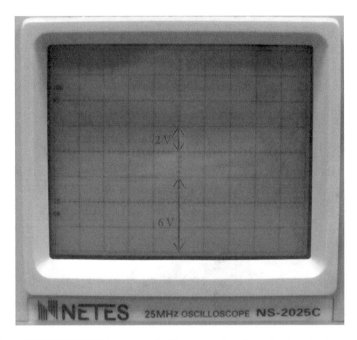

Figure 4.37: When VOLTS/DIV is set to 2 V, one unit of vertical axis shows 2 V and 3 units of vertical axis show 6 V.

Figure 4.38: TIME/DIV control of NETES NS-2025C model analog oscilloscope.

Figure 4.39: TIME/DIV control of GW INSTEK GDS-1022 model digital oscilloscope.

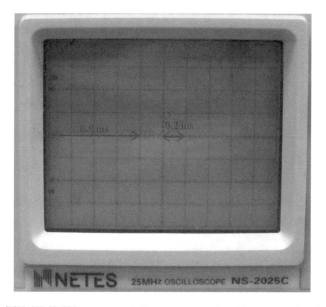

Figure 4.40: When TIME/DIV is set to 0.2 ms, one unit of horizontal axis shows 0.2 ms and 4 units of horizontal axis show 0.8 ms.

Figure 4.41: The value of VOLTS/DIV and TIME/DIV is shown at the bottom of the screen. VOLTS/DIV for channel 1 is 2 V and for channel 2 is 50 mV. TIME/DIV is 500 ns.

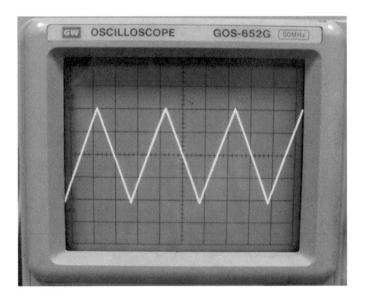

Figure 4.42: A triangular waveform.

Note 1: In digital oscilloscopes, the value of VOLTS/DIV for each channel is shown at the bottom of the screen (see Fig. 4.41).

Example 4.1 Assume that you saw the waveform shown in Fig. 4.42 on CRT of an analog oscilloscope. The VOLT/DIV is set to 0.5 V and TIME/DIV is set to 0.2 ms. Find the amplitude and frequency of the signal. Assume that the oscilloscope probe is in the x10 mode.

Solution: According to the given information, one unit of x axis is 0.2 ms and one unit of y axis is 0.5 V. Since one period of the function is 3 units, then the period is 3×0.2 ms $= 0.6$ ms and frequency is $\frac{1}{0.6 \text{ ms}} = 1.67$ kHz (see Fig. 4.43).

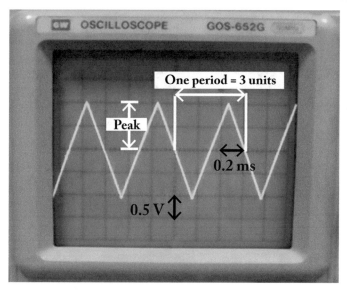

Figure 4.43: Calculation of peak and period.

Peak value of the signal is 2 vertical units. Since VOLT/DIV is 0.5 V, the peak of the signal is $0.5 \times 2 = 1$ V (see Fig. 4.43). So, the signal which entered to the oscilloscope, is 2 Vpp with frequency of 1.67 kHz. Since the signal is attenuated by factor of 10 by the probe, then the original signal which entered to the probe is 10 times bigger. So, the signal which probe measures is a triangular signal with peak to peak of 20 V and frequency of 1.67 kHz.

4.8   COUPLING

All the oscilloscopes permit the user to select among three types of couplings: GND, AC, and DC. The coupling type selector of an analog oscilloscope is shown in Fig. 4.44. In digital oscilloscopes, there is no coupling selector in the front panel of oscilloscope. The coupling selector is accessible in the oscilloscope menus, i.e., digital oscilloscopes use soft coupling selector.

The GND type of coupling is for setting the zero line, i.e., the level for 0 V, of the oscilloscope. Put the coupling in the GND mode and use the POSITION controls in order to place the shown horizontal line on the center of screen, i.e., the x axis. The center of the screen is the best position for setting the ground line of an oscilloscope (Fig. 4.45). Because you have the maximum swing for the positive and negative cycles of input.

Assume a signal like:

$$v(t) = V_0 + V_m \sin(\omega t + \varphi).$$

Figure 4.44: **NETES NS-2025C** model analog oscilloscope coupling selector.

Figure 4.45: The red line is the best position for setting the GND line of an oscilloscope.

V_0 which has the zero frequency is called the DC component (or DC value or average value) of signal. When you exclude the DC term from the signal, the remaining part is called the AC component of signal.

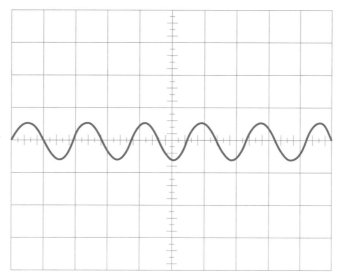

Figure 4.46: What you see with AC coupling.

If you put the coupling selector in the AC mode, then the oscilloscope will filter the DC component, i.e., show the AC component only. So what is shown on the screen is $V_m \sin(\omega t + \varphi)$ (see Fig. 4.46).

If you put the coupling selector in the DC mode, then the oscilloscope doesn't filter the DC component. So in the DC coupling mode, all the components (both DC component and AC component) will be shown (see Fig. 4.47). Figure 4.48 is a simple diagram to understand the different types of coupling.

Assume you want to measure the ripple of a power supply you have in the laboratory. Let's assume that the power supply set to 12 V. The ripple of laboratory power supplies is in the millivolt range. If you put the coupling selector in the DC state, then you can't see the ripple component at all. Because the DC value (12 V) is much bigger than the ripple component and using the DC coupling, you see only a horizontal line at 12 V on the screen of oscilloscope. If you select the AC coupling, the DC value of input signal will be filtered and only the AC component (ripple) will appear on the screen. By selecting a small enough value in the millivolt range for VOLTS/DIV, you can see and measure the ripple component.

4.9   TRIGGER

An oscilloscope's trigger function is important to achieve clear signal characterization, as it synchronizes the horizontal sweep of the oscilloscope to the proper point of the signal. The trigger control enables users to stabilize repetitive waveforms as well as capture single-shot waveforms.

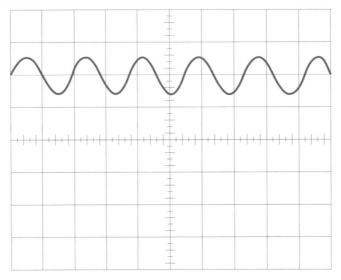

Figure 4.47: What you see with DC coupling.

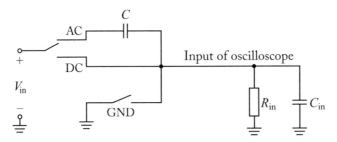

Figure 4.48: A simple diagram to show different types of couplings. The capacitor shown in the AC state, blocks the DC term however acts as a short circuit for the AC components of signal.

By repeatedly displaying similar portion of the input signal, the trigger makes repetitive waveform look static.

Oscilloscopes offer various types of trigger functions, with edge triggering is the most basic and common type. Just like edge triggering, threshold triggering, is another type of trigger function that is offered both in analog and digital oscilloscopes (Figs. 4.49 and 4.50).

When you don't have a stable waveform on the screen, change the trigger LEVEL until it will stabilized. There are many different triggering techniques and each technique have its own application area, however edge triggering is enough for most of applications. In order to obtain more information about different triggering techniques, refer to references introduced at the end of this chapter.

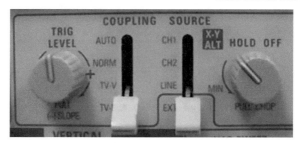

Figure 4.49: Trigger section of NETES NS-2025C model analog oscilloscope.

Figure 4.50: Trigger section of GW INSTEK GDS-1022 model digital oscilloscope.

4.10   SAMPLE OSCILLOSCOPE

GW INSTEK GDS-1022 model digital oscilloscope is shown in Fig. 4.51. In order to turn on the device, push the "POWER" button (Fig. 4.52).

The unit of horizontal axis is set with TIME/DIV knob and the unit of vertical axis is set with VOLTS/DIV knob (Fig. 4.53). Turn these knobs in order to obtain the desired TIME/DIV and VOLTS/DIV. Selected values of TIME/DIV and VOLTS/DIV are shown on the screen (Fig. 4.54).

Figure 4.51: GW INSTEK GDS-1022 model oscilloscope.

Figure 4.52: Power button.

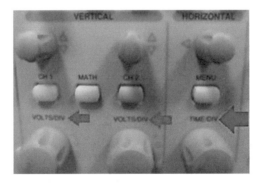

Figure 4.53: Each channel has its own VOLTS/DIV control. There is only one TIME/DIV control.

Figure 4.54: VOLTS/DIV for channel 1 is 2 V and for channel 2 is 50 mV. Value of TIME/DIV (for both channels) is 500 ns.

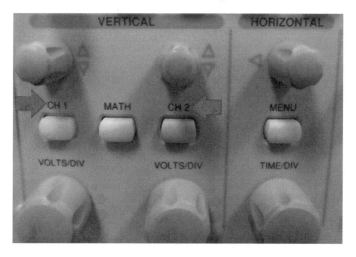

Figure 4.55: CH 1 and CH 2 buttons.

4.11   TYPE OF COUPLING

You can select the desired type of coupling with the aid of CH 1/CH 2 buttons (see Fig. 4.55). For instance, assume that you want to set the channel 2 coupling to your desired value. In order to do this, press the CH 2 button. After CH 2 is pressed, a menu will appear in the right half of the screen (see Fig. 4.56). Press the F1 button in order to change the coupling and select the desired on. Note that the type of coupling is shown with the aid of icons shown in Figs. 4.57–4.59.

The type of coupling of each channel is shown on the left side of the screen (Fig. 4.60).

Note: You can show/hide a signal with the aid of CH 1/CH 2 buttons as well. When you press the CH 1/CH 2 buttons, a menu will appear on the right side of the screen (see Fig. 4.56). If you press CH 1/CH 2 once again, then the waveform will be disappeared from the screen. You can press the CH 1/CH 2 buttons again in order to see the hidden waveform.

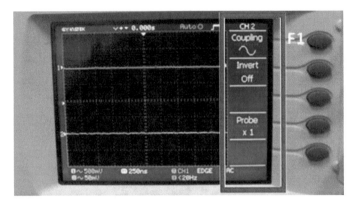

Figure 4.56: Type of coupling is set with the F1 button.

Figure 4.57: Icon for AC coupling.

Figure 4.58: Icon for GND coupling.

Figure 4.59: Icon for DC coupling.

Figure 4.60: Coupling of channel 1 and 2 is DC.

Figure 4.61: CH 1 button.

4.12 PROCEDURE TO SEE THE WAVEFORM ON THE SCREEN

4.12.1 MANUAL METHOD

In order to see a waveform with channel 1, do the following.

1. Press "CH 1" button (see Fig. 4.61) and set the coupling of channel 1 to ground (GND).

2. Use the channel 1 vertical control (see Fig. 4.62) to put the horizontal line of channel 1 on the center of the screen, i.e., x axis of the coordinate shown on the screen (see Fig. 4.63).

3. Select the suitable type of coupling based on the type of signal that you want to see.

4. Select X1 or X10 of probe based on the amplitude of the waveform that you want to see. If you want to measure a big signal (For instance, a signal with a peak value of bigger than 20 V), it is a good idea to decrease its amplitude before entering it into the scope. If you want to measure a small signal, then put the probe in the X1 mode. After setting the probe in the suitable state, you need to announce it to the oscilloscope. In order to do this, press the CH 1 button and use the F4 button in order to set the Probe (see Fig. 4.64) to the same state as you have on the probe. If your probe switch is set to X1, then you need to set the Probe to X1. If you set the probe switch to X10, then you need to set Probe to X10.

5. Connect the probe to the points that you want to see its waveform. Use the VOLTS/DIV and TIME/DIV in order to see a clear image of waveform. You need to see at least one full period of the waveform on the screen.

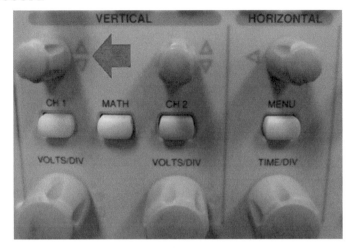

Figure 4.62: Channel 1 vertical control.

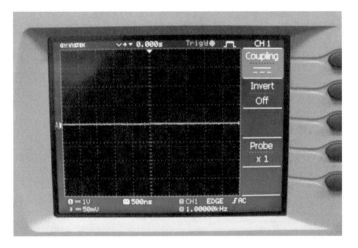

Figure 4.63: The ground line of channel 1 is set to center of screen.

Note 1:   It is a good idea to test your oscilloscope and ensure that it measures correctly before starting your measurement. In order to test your oscilloscope, you can connect the tip of probe to the test point on the front panel of the oscilloscope. The test signal is a square wave with a frequency of 1 kHz. However, the amplitude of this square wave may change from oscilloscope to oscilloscope. For the GW INSTEK GDS-1022, the peak to peak of this square wave is about 2 V (see Fig. 4.65). When you a see a square with peak to peak of 2 V and frequency of 1 kHz (see Fig. 4.66), then your oscilloscope is ready for measurement.

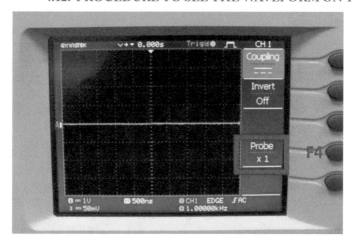

Figure 4.64: Entering the oscilloscope probe switch status.

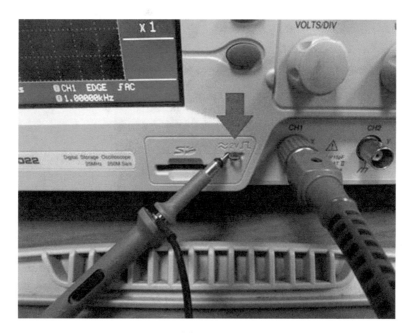

Figure 4.65: Calibration point of GW INSTEK GDS-1022.

Note 2:   If the waveform moves on the screen and you don't have a stable image of it, then rotate the TRIGGER LEVEL control in order to obtain a stable waveform (Fig. 4.67).

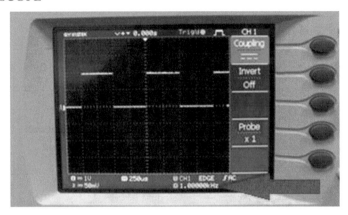

Figure 4.66: Frequency of periodic signals are shown in the bottom of the screen.

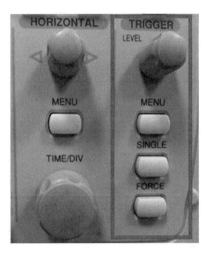

Figure 4.67: Use the LEVEL knob to stabilize a moving waveform.

6. After you see the stable waveform on the screen, you can do the measurements that you want. As shown in Fig. 4.68, the settling time of the square waveform is 2.2 μs and as shown in Fig. 4.69, the peak of the waveform is 2 V.

4.12.2 AUTOMATIC METHOD

Digital oscilloscopes could select the best settings for you automatically. In order to use this capability of digital oscilloscopes,

1. connect the probe(s) to the circuit under test; and

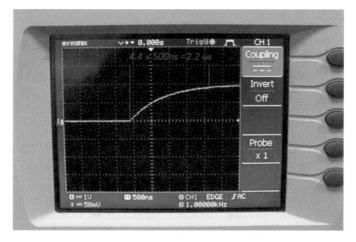

Figure 4.68: You need to decrease the TIME/DIV in order to see the transients.

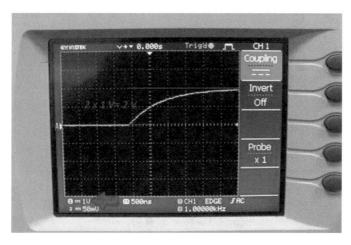

Figure 4.69: Peak of signal is 2 V. VOLTS/DIV is set to 1 V.

2. press the "Autoset" button. The waveforms of both CH 1 and CH 2 will be appear after few seconds on the screen (Fig. 4.71).

4.13   AUTOMATIC MEASUREMENT WITH DIGITAL OSCILLOSCOPES

After you see a stable waveform on the screen, you can use the Measure button (see Fig. 4.72) in order to do measurement on the shown waveform.

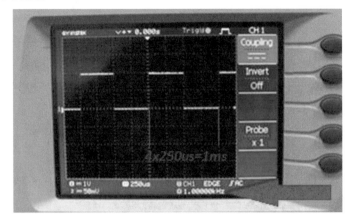

Figure 4.70: Period of signal is 1 ms. The frequency of signal is 1 kHz.

Figure 4.71: Autoset button could show the waveform on the screen automatically.

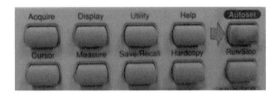

Figure 4.72: Measure button.

After pressing the Measure button, a menu will be appearing on the right side of the screen. Using the F1, F2, F3, F4, and F5 buttons you can determine the type of appeared measurements (Fig. 4.73).

After you pressed one of the F1, F2, F3, F4, and F5 buttons, the "Select Measurement" window will appear (Fig. 4.74).

Use the VARIABLE knob (see Fig. 4.75) in order to set the desired measurement.

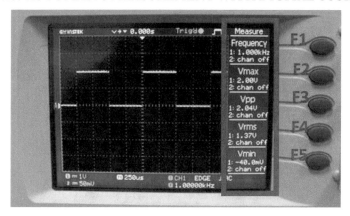

Figure 4.73: Measured values are shown on the right side of the screen.

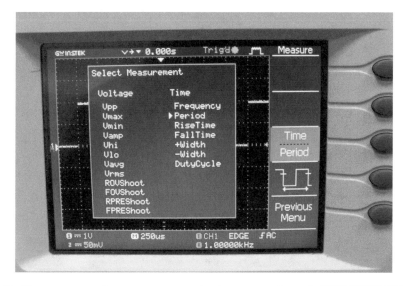

Figure 4.74: Different types of measurements which you can do with GW INSTEK GDS-1022.

After you select the desired measurement, press the button corresponding to "Previous Menu" (F5 button) shown in Fig. 4.76.

Note 1: Before using the automatic measurement, ensure that at least one full period of signal is shown on the screen.

Note 2: Don't forget to announce the states of probe switches correctly to the scope, otherwise the voltage based measurement for instance RMS, peak to peak, mean value measurements, etc., will not be done correctly.

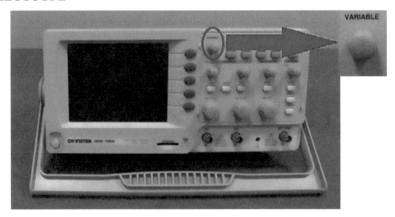

Figure 4.75: Variable knob.

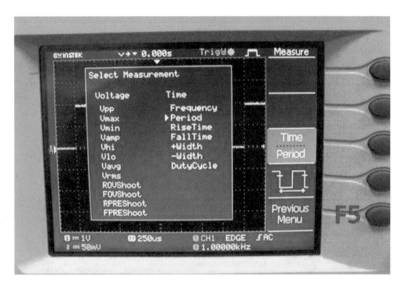

Figure 4.76: Press F5 in order to return to the previous page.

4.14   CURSORS

Cursors are one of the most important tools of digital oscilloscopes. With the aid of cursors, you can read time or voltages very easily.

Assume we are seeing the waveform shown in Fig. 4.77 on the screen. We want to use cursors in order to do some measurements, for instance assume we want to measure the settling time. In order to activate the cursors, press the Cursor button (see Fig. 4.78).

After the Cursor button is pressed, two cursors will be added to the screen (see Fig. 4.79). With the aid of these two vertical cursors you can do the time measurements.

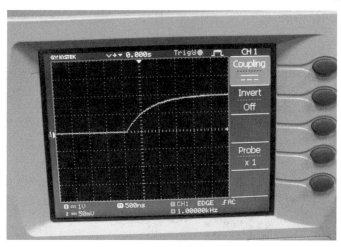

Figure 4.77: Manual measurement of settling time with cursors.

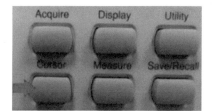

Figure 4.78: Cursor button.

Note: With the aid of the F1 button (see Fig. 4.80), you can select the source of measurement. If you select CH 1, then you can measure the waveform of channel 1 and if you select the CH 2, you can measure the waveform of channel 2.

In order to activate the first cursor, press the F2 button (see Fig. 4.81). After pressing the F2 button, first cursor will be activated and you can move it to where you want with the aid of VARIABLE knob (see Fig. 4.75).

Move the first cursor. Note that the coordinate of the cursor appears on the right side of the screen (Fig. 4.82).

Press F3 in order to activate the second cursor. Like the first cursor, you can move this cursor with the aid of the VARIABLE knob. Put this cursor wherever you want. The difference between the coordinate of the first and second cursors are shown in the X1X2 section (see Fig. 4.83) on the right of the screen.

Note: If you want to do vertical (voltage) measurements, then it is better to use horizontal cursors. To do this, press the button corresponding to X ↔ Y (F5 button). Activate and bring

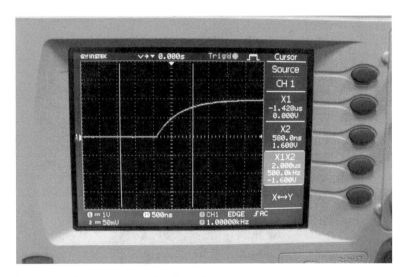

Figure 4.79: Two vertical cursors are added to the screen.

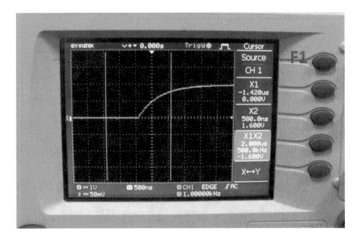

Figure 4.80: Use the F1 button to select the source of measurement.

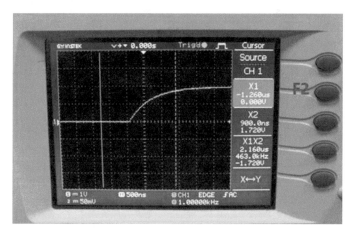

Figure 4.81: Activate the first cursor with F2 button.

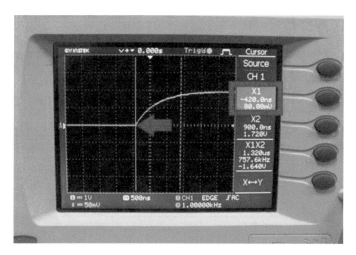

Figure 4.82: As you move the cursor, the coordinate of the cursor appears on the screen.

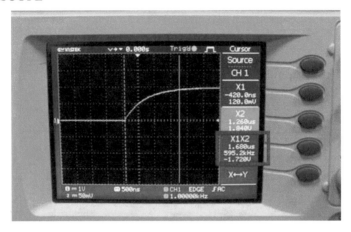

Figure 4.83: Time and voltage difference between the points which intersect the two cursors. The intersections are shown with red stars.

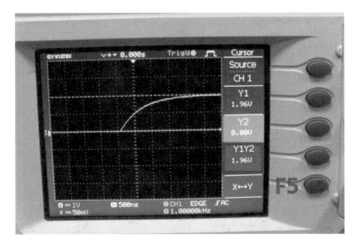

Figure 4.84: Voltage measurement with horizontal cursors.

the horizontal cursors where you want them in the same way that you did for vertical cursors (Fig. 4.84).

4.15   PHASE DIFFERENCE MEASUREMENT

You can use two different methods in order to measure the phase difference between two sinusoidal waveforms.

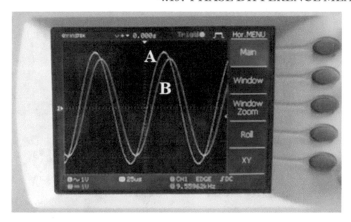

Figure 4.85: There is a phase difference between A and B.

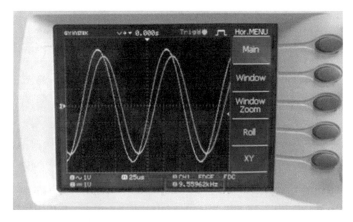

Figure 4.86: Frequency of waveforms are shown in the bottom of the screen.

4.15.1 FIRST METHOD: MEASUREMENT OF DELAY BETWEEN THE TWO WAVEFORMS

Assume that you want to measure the phase difference between the two waveform shown in Fig. 4.85.

In order to do this,

1. calculate the period of the signal. According to Fig. 4.86, the frequency is 9.56 kHz so the period is $\frac{1}{9.56 \text{ kHz}} = 104.6 \ \mu s$;

2. measure the delay between the two signals. We use the cursors for this purpose. According to Fig. 4.87, the delay between the two signal is 8.2 μs; and

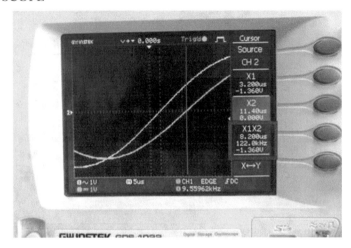

Figure 4.87: Measurement of time delay between the two waveforms.

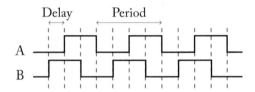

Figure 4.88: Phase difference between two square wave A and B is: $\Delta\varphi = \frac{\text{Delay}}{\text{Period}} \times 360°$.

3. use the $\Delta\varphi = \frac{\text{Delay}}{\text{Period}} \times 360°$ relation in order to calculate the phase difference between the two waveforms. After putting the numbers into the aforementioned equation, results to $\frac{8.2 \ \mu s}{104.6 \ \mu s} \times 360° = 28.23°$.

Note: You can use this method in order to measure the phase difference between two square waves as well (Fig. 4.88).

4.15.2 SECOND METHOD: LISSAJOUS CURVE METHOD

You can use the Lissajous curve in order to measure the phase difference. In order to measure the phase difference between the two waveforms using the Lissajous curves, do the following.

1. Press the "HORIZONTAL MENU" button (see Fig. 4.89). A menu will appear on the right side of the screen (see Fig. 4.90). Press F5 in order to select the "XY" mode. After you press F5, the Lissajous curve will be appear on the screen.

2. Phase difference could be calculated using the $\Delta\varphi = \sin^{-1}(\frac{B}{A})$ equation. A and B could be defined as shown in Fig. 4.91 or Fig. 4.92. The values of A and B of Fig. 4.91 are not

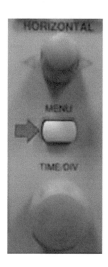

Figure 4.89: Menu button.

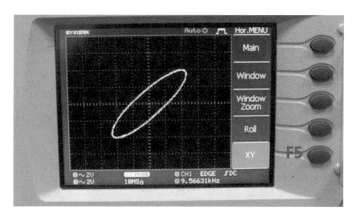

Figure 4.90: Lissajous curve.

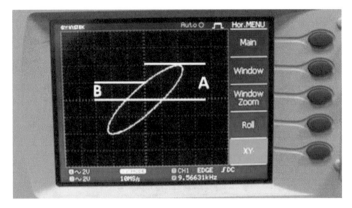

Figure 4.91: Measurements of A and B.

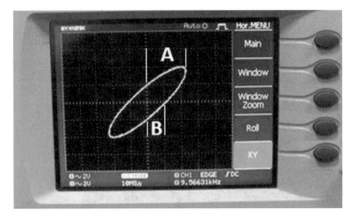

Figure 4.92: Measurements of A and B.

necessarily the same as the A and B of Fig. 4.92. However, the ratio $\frac{B}{A}$ of both figures is the same.

You can use cursors in order to measure the values of B and A. According to Fig. 4.93, B = 2.00 V and A = 3.92 V. So,

$$\Delta\varphi = \sin^{-1}\left(\frac{2}{3.92}\right) = 30.67°.$$

You use the vertical cursors and measure the A and B as defined in Fig. 4.94.

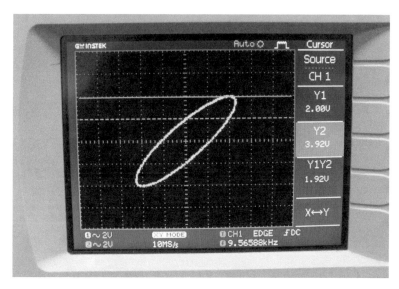

Figure 4.93: A = 3.92 V and B = 2 V.

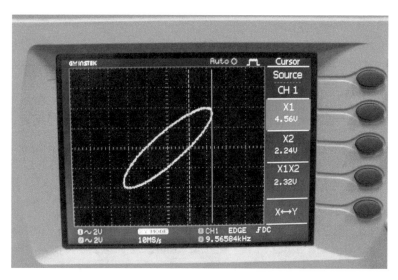

Figure 4.94: A = 4.56 V and B = 2.24 V.

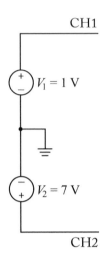

Figure 4.95: Simple series connection of two DC sources.

4.16   ADDITION/SUBTRACTION OF CHANNEL WAVEFORMS

You can generate the addition/subtraction of waveforms entered to the oscilloscope channels. We show this capability with an example. Assume a circuit like the one shown in Fig. 4.95. Realization of this circuit is shown in Fig. 4.96.

In order to see the addition of the two waveforms shown on the screen (Fig. 4.97),

1. press the "MATH" button (see Fig. 4.98). After you pressed the MATH button, a menu will appear on the right side of the screen (see Fig. 4.99);

2. use "F1" in order to select the "Operation CH 1 + CH 2" (Fig. 4.100); and

3. the addition of two channels will appear on the screen with red color. In this oscilloscope, the VOLTS/DIV for the red waveform is the same as the VOLTS/DIV of CH 2. So, according to Fig. 4.101, the red waveform shows 1.6×5 V $= 8$ V. This value is expected since $1 + 7 = 8$.

If you select the "OPERATIN CH 1 − CH 2," then the subtraction of two waveforms will appeared on the screen. According to Fig. 4.102, difference of these two waveforms is -1.2×5 V $= -6$ V. This value is expected since $1 - 7 = -6$.

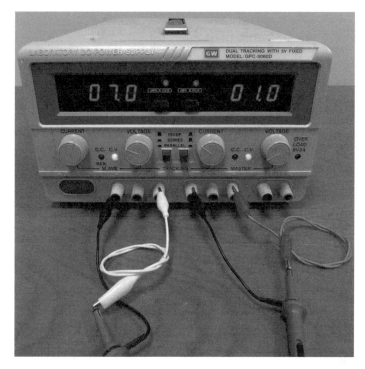

Figure 4.96: Realization of Fig. 4.95.

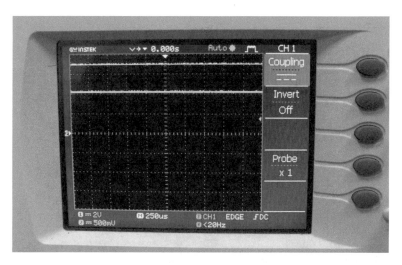

Figure 4.97: Channel 1 is 7 V and Channel 2 is 1 V.

Figure 4.98: Math button.

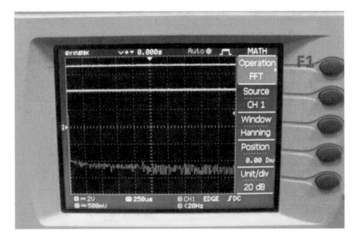

Figure 4.99: Math menu.

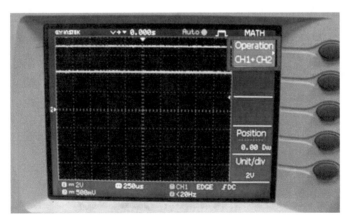

Figure 4.100: Selection of "Operation CH 1 + CH 2."

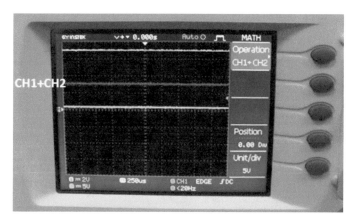

Figure 4.101: Result of the addition of two channels is shown in red.

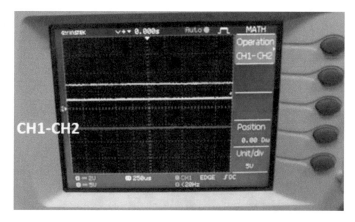

Figure 4.102: The difference between channel 1 and 2 is shown in red.

4.17   FURTHER READING

[1] David Herres, *Oscilloscopes: A Manual for Students, Engineers, and Scientists*, Springer, 2020. DOI: 10.1007/978-3-030-53885-9.

[2] Roman Malaric, *Instrumentation and Measurement in Electrical Engineering*, Brown Walker Press, 2011.

Refer to the following references in order to learn more about different triggering methods:

[3] Triggering Fundamentals. https://download.tek.com/document/55W_17291_6_0.pdf

[4] XYZs of Oscilloscopes. https://www.tek.com/document/online/primer/xyzs-scopes/ch4/oscilloscope-systems-and-controls

[5] https://www.picotech.com/library/oscilloscopes/advanced-digital-triggers

CHAPTER 5

Drawing the Graph of Data with MATLAB®

5.1 INTRODUCTION

Drawing the graph of data is an important skill to have. You can get a lot of information by looking at a graph. For instance, you can see whether or not the changes are linear. You can see the effect of increasing/decreasing the independent variable, etc.

In this section, we will show you how you can visualize your data using MATLAB.

5.2 DRAWING GRAPH OF MEASURED DATA

Assume that we obtained the data shown in Table 5.1. This data shows the voltage and current through a resistor.

First of all, we need to enter the data into MATLAB (Fig. 5.1).

Table 5.1: Voltage and current for a resistor

V (Volt)	I (Amper)
0.499	0.10
0.985	0.20
1.508	0.31
1.969	0.41
2.528	0.53
2.935	0.61
3.481	0.73
3.971	0.83
4.486	0.94
4.960	1.04
5.502	1.15
6.007	1.26
6.600	1.38

```
Command Window                                                                    ⊙
  >> V=[.499 .985 1.508 1.969 2.528 2.935 3.481 3.971 4.486 4.960 5.502 6.007 6.600];
  >> I=[.1 .2 .31 .41 .53 .61 .73 .83 .94 1.04 1.15 1.26 1.38];
fx >>
```

Figure 5.1: Entering the Table 5.1 data into MATLAB.

```
Command Window                                                                    ⊙
  >> V=[.499 .985 1.508 1.969 2.528 2.935 3.481 3.971 4.486 4.960 5.502 6.007 6.600];
  >> I=[.1 .2 .31 .41 .53 .61 .73 .83 .94 1.04 1.15 1.26 1.38];
  >> plot(V,I),grid on
fx >>
```

Figure 5.2: Drawing the graph of Table 5.1 data.

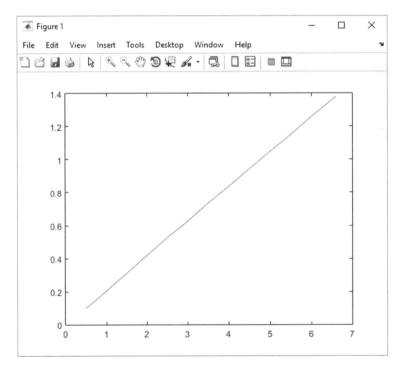

Figure 5.3: Graph of Table 5.1 data.

You can use the "plot" command in order to draw the plot of data. Use the "grid" command in order to add grids to your graph (Figs. 5.2 and 5.3).

Use the "grid minor" if you prefer smaller grid lines (Fig. 5.4).

```
Command Window                                                              ⊙
  >> V=[.499 .985 1.508 1.969 2.528 2.935 3.481 3.971 4.486 4.960 5.502 6.007 6.600];
  >> I=[.1 .2 .31 .41 .53 .61 .73 .83 .94 1.04 1.15 1.26 1.38];
  >> plot(V,I),grid on
  >> plot(V,I),grid minor
fx >>
```

Figure 5.4: The `grid minor` command decreases the size of grids.

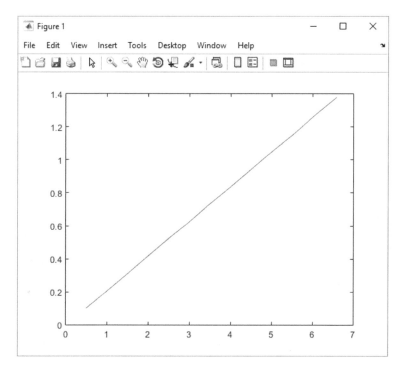

Figure 5.5: Addition of smaller grid to the graph.

Data points are not shown in the graphs of Figs. 5.3 and 5.5. Using the commands shown in Fig. 5.6, you can show the data points with a red star (Fig. 5.7).

A graph without a label is worth nothing. So, always be careful to add label to your graphs. This could be done with the aid of "xlabel" and "ylabel" commands (Figs. 5.8 and 5.9).

Instead of xlabel and ylabel commands, you could use the Insert menu in order to add label to your axis (Fig. 5.10).

You can save the obtained graph with bmp format as well. In order to do this, use the **File>Save As...** (see Fig. 5.11). After that select the bmp format from the Save as Type drop down list (Fig. 5.12).

```
Command Window
  >> V=[.499 .985 1.508 1.969 2.528 2.935 3.481 3.971 4.486 4.960 5.502 6.007 6.600];
  >> I=[.1 .2 .31 .41 .53 .61 .73 .83 .94 1.04 1.15 1.26 1.38];
  >> plot(V,I),grid on
  >> plot(V,I),grid minor
  >> plot(V,I,'r*',V,I),grid minor
fx >>
```

Figure 5.6: Commands required for showing the data points with a red star.

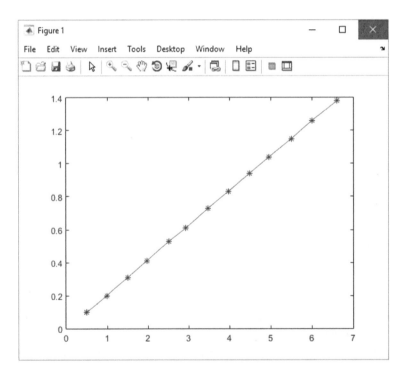

Figure 5.7: Data points are shown with a red star.

```
Command Window                                                    ⊙
  >> V=[.499 .985 1.508 1.969 2.528 2.935 3.481 3.971 4.486 4.960 5.502 6.007 6.600];
  >> I=[.1 .2 .31 .41 .53 .61 .73 .83 .94 1.04 1.15 1.26 1.38];
  >> plot(V,I),grid on
  >> plot(V,I),grid minor
  >> plot(V,I,'r*',V,I),grid minor
  >> xlabel('Resistor Voltage(V)')
  >> ylabel('Resistor Current(A)')
fx >>
```

Figure 5.8: Addition of label to the graph with the aid of xlabel and ylabel commands.

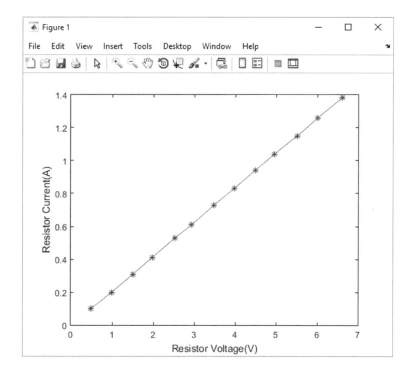

Figure 5.9: Axes are labeled.

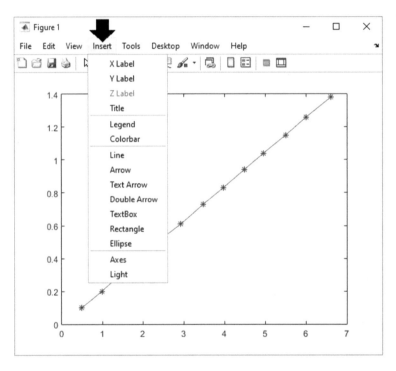

Figure 5.10: Labels can be added to the graph with the aid of the Insert menu.

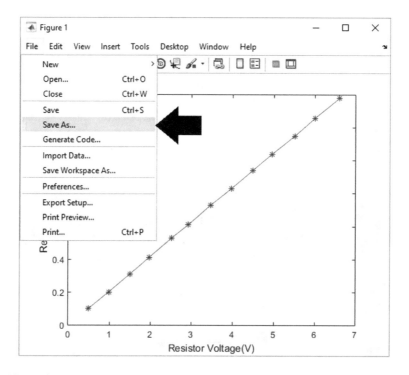

Figure 5.11: "Save As…" could be used to export the graph as a .bmp file.

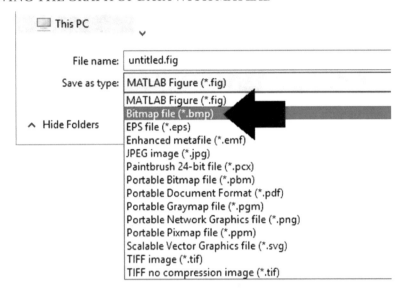

Figure 5.12: Selection of .bmp output file.

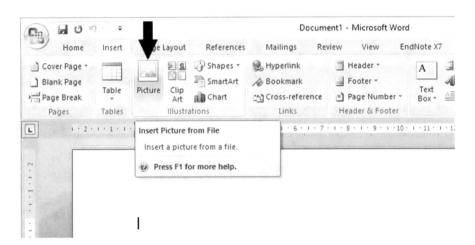

Figure 5.13: Picture button of Microsoft Word.

You can use the Insert menu of Microsoft Word in order to add a picture to your report (Fig. 5.13).

Table 5.2: Voltage and current for a resistor with nominal value of 5.6 Ω

V (Volt)	I (Amper)
0.579	0.10
0.978	0.17
1.598	0.28
1.976	0.34
2.496	0.43
2.953	0.51
3.458	0.60
4.068	0.71
4.450	0.78
4.917	0.86
5.350	0.93
5.750	1.01
6.370	1.11
6.600	1.15

```
Command Window                                                              ⊙
  >> I1=[.1 .2 .31 .41 .53 .61 .73 .83 .94 1.04 1.15 1.26 1.38];
  >> V1=[.499 .985 1.508 1.969 2.528 2.935 3.481 3.971 4.486 4.960 5.502 6.007 6.600];
  >> I2=[.1 .17 .28 .34 .43 .51 .60 .71 .78 .86 .93 1.01 1.11 1.15];
  >> V2=[.579 .978 1.598 1.976 2.496 2.953 3.458 4.068 4.450 4.917 5.35 5.75 6.37 6.6];
  >> plot(V1,I1,'*r',V1,I1)
  >> hold on
  >> plot(V2,I2,'+r',V2,I2)
  >> grid minor
fx >> |
```

Figure 5.14: Entering the data to MATLAB.

5.3   DRAWING TWO GRAPH ON THE SAME AXIS

Table 5.2 shows the voltage and current measurement for a resistor with nominal value of 5.6 Ω. We want to draw the graph of this data on the same axis as the one shown on the graph of Table 5.1. Drawing on the same axis permits us to compare different measurements.

In order to do this, use the following code (see Fig. 5.14). The data points of Table 5.2 are shown with red + signs (Fig. 5.15).

Use the **Insert>Legend** in order to add a guide to your plot (Figs. 5.16 and 5.17).

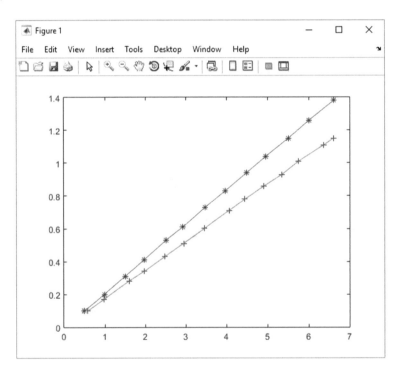

Figure 5.15: Data for Tables 5.1 and 5.2 are shown on the same graph.

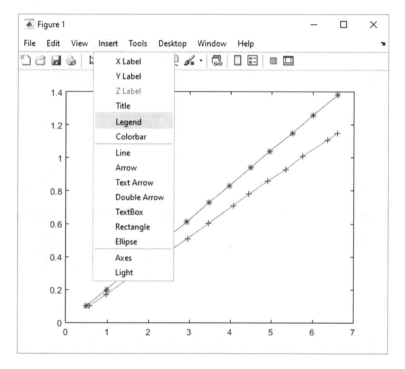

Figure 5.16: Addition of legend to the graph.

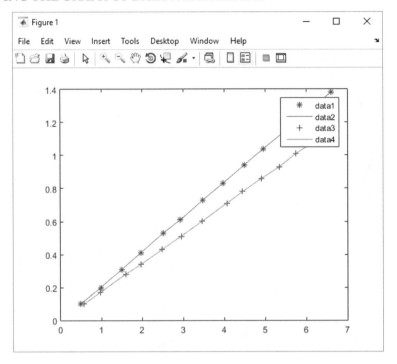

Figure 5.17: A legend is added to the graph. You can click on the legend and change its position.

Double click the "data1," "data2," "data3," and "data4" and change them to what you want (Fig. 5.18).

5.4   DRAWING FREQUENCY RESPONSE GRAPHS

Assume we want to draw the graph of following frequency response.

Table 5.3 is the frequency response of following simple RC circuit (Fig. 5.19). First, we need to enter the data into MATLAB (Fig. 5.20). Use the `semilogx` command in order to draw the amplitude and phase graph of the given data (Fig. 5.21). Use the **Insert>Title** in order to add a title to the graph. You can add a label to the vertical and horizontal axes as well (Figs. 5.23 and 5.24).

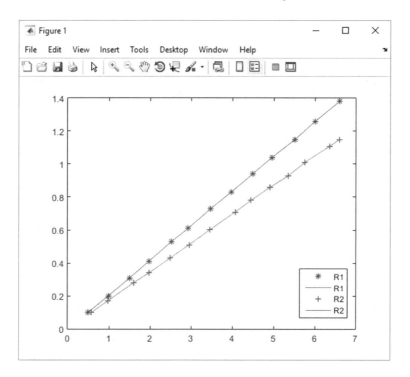

Figure 5.18: You can change the legend text to what you want.

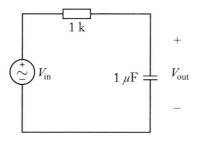

Figure 5.19: A simple RC circuit.

```
Command Window
>> f=[1 10 20 50 100 150 200 250 300 350 400 450 500 550 600];
>> g=[1 .998 .992 .954 .847 .728 .623 .537 .469 .414 .370 .333 .303 .278 .256];
>> faz=[-.36 -3.6 -7.16 -17.44 -32.13 -43.30 -51.48 -57.51 -62.05 -65.54 -68.30 -70.51 -72.34 -73.85 -75.14];
fx >>
```

Figure 5.20: Entering the frequency response to MATLAB.

Table 5.3: Frequency response of simple RC circuit

Frequency (Hz)	Gain $\left(\left\|\dfrac{V_o(jw)}{V_{in}(jw)}\right\|\right)$	Phase $\left(\measuredangle\dfrac{V_o(jw)}{V_{in}(jw)}\right)$ degrees
1	1.000	−0.36
10	0.998	−3.60
20	0.992	−7.16
50	0.954	−17.44
100	0.847	−32.13
150	0.728	−43.30
200	0.623	−51.48
250	0.537	−57.51
300	0.469	−62.05
350	0.414	−65.54
400	0.370	−68.30
450	0.333	−70.51
500	0.303	−72.34
550	0.278	−73.85
600	0.256	−75.14

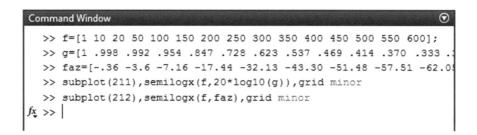

```
Command Window
   >> f=[1 10 20 50 100 150 200 250 300 350 400 450 500 550 600];
   >> g=[1 .998 .992 .954 .847 .728 .623 .537 .469 .414 .370 .333 .
   >> faz=[-.36 -3.6 -7.16 -17.44 -32.13 -43.30 -51.48 -57.51 -62.0
   >> subplot(211),semilogx(f,20*log10(g)),grid minor
   >> subplot(212),semilogx(f,faz),grid minor
fx >> |
```

Figure 5.21: Commands required to draw the frequency response plot (Fig. 5.22). Variable f is frequency, g is the magnitude, and faz is the phase.

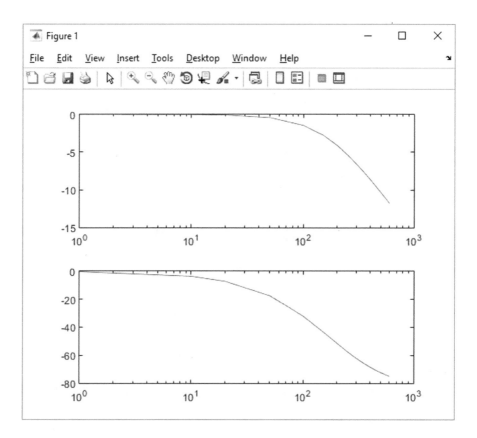

Figure 5.22: Frequency response plot of Table 5.3.

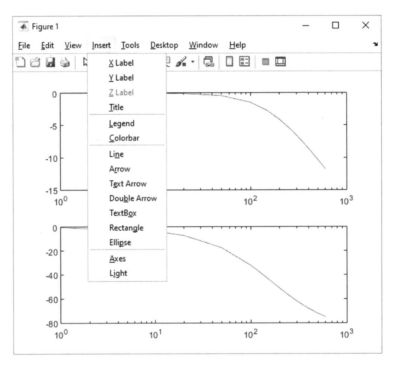

Figure 5.23: Addition of title to the graph.

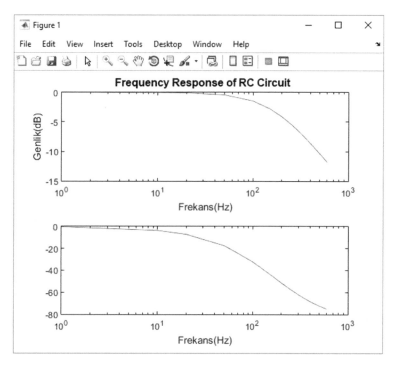

Figure 5.24: A title is added to the graph.

5.5   FURTHER READING

[1] Stephen J. Chapman, *MATLAB Programming for Engineers*, Cengage Learning, 2019.

Authors' Biographies

FARZIN ASADI

Farzin Asadi received his B.Sc. in Electronics Engineering, his M.Sc. in Control Engineering, and his Ph.D. in Mechatronics Engineering. Currently, he is with the Department of Electrical and Electronics Engineering at the Maltepe University, Istanbul, Turkey.

Dr. Asadi has published more than 40 international papers and 13 books. He is on the editorial board of seven scientific journals as well. His research interests include switching converters, control theory, robust control of power electronics converters, and robotics.

KEI EGUCHI

Kei Eguchi received his B.Eng., M.Eng., and D.Eng. degrees from Kumamoto University, Kumamoto, Japan, in 1994, 1996, and 1999, respectively. His research interests include nonlinear dynamical systems, intelligent circuits and systems, and low-voltage analog integrated circuits.

From 1999–2006, he was an Associate Professor and a Lecturer in Kumamoto National College of Technology. From 2006–2012, he was an Associate Professor in Shizuoka University. In 2012, he joined the faculty of Fukuoka Institute of Technology, where he is now a Professor.

Prof. Dr. Eguchi received ICICIC2018 Best Paper Award, IETNR-18 Oral Best Paper Award, ICICIC2017 Best Paper Award, ICICIC2016 Best Paper Award, ICEEI2016 Excellent Oral Presentation Award, ICIAE2016 Best Presentation Award, ICEESE2016 Best Presenter Award, ICIAE2015 Best Presentation Award, ICPEE2014 Excellent Oral Presentation Award, iCABSE2014 Excellent Paper Award, KKU-IENC2014 Outstanding Paper Award,

ICEEN2014 Excellent Paper Award, JTL-AEME2013 Best Paper Award, ICTEEP2013 Best Session Paper Award, 2010 Takayanagi Research Encourage Award, 2010 Paper Award of Japan Society of Technology Education, ICICIC2009 Best Paper Award, and ICINIS2009 Outstanding Contribution Award. He is a senior member of IEEJ and a member of INASS and JSTE.

Printed in the United States
by Baker & Taylor Publisher Services